Praise for *Farming on the Wild Side*

"I love this book, which is an incredible and inspiring ray of hope. *Farming on the Wild Side* shows both experienced and aspiring farmers how to build a highly productive, biodiverse, and profitable family farm *and* have a fulfilling lifestyle. This is the future of farming."

— ANDRÉ LEU, international director, Regeneration International; author of *Poisoning Our Children*

"As a fellow farmer and longtime member of the organic community, the journey Nancy and John Hayden take us on with their new book resonates in my mind and carries me back through similar transitions with my own farm. While the book touches gently on many topics, I appreciate the realistic view of farming and rural life and their very real commitment to the concepts of soil health."

— JEFF MOYER, executive director, Rodale Institute

"With *Farming on the Wild Side*, Nancy and John Hayden have delivered an inspiring story of shepherding an old, tired Vermont dairy into a new, diversified organic farm that serves the needs of the twenty-first century. It covers their journey step-by-step with new crops, new markets, and new farming methods, modeling an ecological balance that farms must achieve in the future. But the book is more than their personal experience with innovation over three decades; it's also a philosophical and practical guide to restoring land to health, which benefits the farmer, one's community, and all living things. Given the pace of climate change and the importance of regenerative agriculture as a key solution, this book is very timely!"

— WILL RAAP, founder and chairman, Gardener's Supply

"This amazing book details how Nancy and John are living the new farming paradigm, one that maximizes ecosystem functionality and values soil, biodiversity, human well-being, and long-term resiliency. Their farming journey and philosophical evolution provide practical and science-based solutions for how the backyard gardener, hobby farmer, or large-scale grower can be part of the pollinator, food, and climate solution."

— HEATHER HOLM, pollinator educator; author of *Bees* and *Pollinators of Native Plants*

"Good farmers learn how to listen to the land. Nancy and John Hayden work with Nature to produce wholesome food for their family and community. Life doesn't get better than this. *Farming on the Wild Side* provides plenty of practical advice and green inspiration to up your growing game."

— MICHAEL PHILLIPS, author of *The Holistic Orchard* and *Mycorrhizal Planet*

"An inspiration—*Farming on the Wild Side* lays down an ecologically justified path for others to follow for a biodiverse farm."

— JO ANN BAUMGARTNER, executive director, Wild Farm Alliance

"This intelligent book is written by two people who have a pure love and appreciation for the land and its inhabitants. *Farming on the Wild Side* serves as an authentic account of the farmers' personal transformation, a practical guide to agroecological transition, and an inspiration to live in greater harmony with nature. The farm that was created and described in detail here truly embodies the concept of a multifunctional landscape, supporting productive, ecological, and cultural functions."

— DR. SARAH LOVELL, H.E. Garrett Endowed Chair Professor and director of the Center for Agroforestry, University of Missouri

"Nancy and John use a forensic but accessible approach to explore their personal journey from scientists to farmers. New crops and enterprises are approached with careful planning, and their successes, or otherwise, are carefully observed to inform future decisions. Even non-farmers will enjoy reading about how they gradually adapted their farm to create a diverse farming system suited to their character and their geography. *Farming on the Wild Side* is a masterclass in working with nature to create abundance."

—BEN RASKIN, Head of Horticulture, Soil Association

"In telling the history of a farm and its farmers, *Farming on the Wild Side* addresses issues of great relevance to the future of agriculture. John and Nancy Hayden remind us that an ecologically based and socially just agriculture needs to involve deep and diverse relationships between people and landscapes. The Haydens present a true example of co-evolution between the Farm Between and its stewards, documenting their use of agroecological principles to transform a conventional dairy into a diversified farm, which takes full advantage of its ecological processes. Each one of the stages of transformation—from dairy to organic vegetables to a regenerative fruit farm and nursery—offer valuable lessons to reflect on. I have been working on research and education with the Haydens for a decade, and they have generously taught me and my students how to fully integrate the science and practice dimensions of agroecology."

—V. ERNESTO MÉNDEZ, PhD, professor of agroecology and environmental studies, University of Vermont

"It's no secret that the way in which we feed ourselves and inhabit the land must change, but few of us truly know how to make that happen. The beauty of this book is that John and Nancy Hayden *do* know; even better, they've graciously shared their knowledge in these pages."

—BEN HEWITT, author of *The Nourishing Homestead*

"I love this book. It takes me back to a time and a place my grandparents used to talk about—a good time and place—when farmers and growers worked with nature instead of against it. *Farming on the Wild Side* inspires, informs, and fills me with hope that we can heal our relationship with the wild. Just reading this book is healing, and I can't wait to put some of Nancy and John Hayden's ways into practice on our own little plot of land."

—BRIGIT STRAWBRIDGE HOWARD, author of *Dancing with Bees*

"This lavishly illustrated book follows the Hayden family as they bring nature back to their farm, and in the process, it tells a story of learning, testing, observing, and creating an agroecological model of how to farm with nature, not against her. They combine the science of biodiversity management, the practices of good farming, and the transformative change humans need in order to return to the kinds of food systems that will feed the land as well as feed us."

—STEVE GLIESSMAN, professor emeritus of agroecology, UC Santa Cruz

"In easy, conversational prose, Nancy and John Hayden offer the aspiring regenerative farmer a compendium of wisdom on the practicalities of establishing, developing, surviving, enjoying, and profiting from the small farm without losing sight of bigger ecological and political issues. Their warts-and-all history of their own farming practice rings true and is full of inspiration for those seeking a better future while dealing with present realities—which is hopefully all of us. We need more books like this."

—CHRIS SMAJE, writer, Small Farm Future; farmer, Somerset, UK

FARMING ON THE WILD SIDE

The Evolution of a Regenerative Organic Farm and Nursery

NANCY J. HAYDEN *and* **JOHN P. HAYDEN**

Chelsea Green Publishing
White River Junction, Vermont
London, UK

Copyright © 2019 by Nancy J. Hayden and John P. Hayden
All rights reserved.

Unless otherwise noted, all photographs by Nancy J. Hayden and John P. Hayden

Illustrations copyright © 2019 by Elara Tanguy

No part of this book may be transmitted or reproduced in any form by any means without permission in writing from the publisher.

Project Manager: Alexander Bullett
Acquisitions Editor: Fern Marshall Bradley
Developmental Editor: Michael Metivier
Copy Editor: Eliani Torres
Proofreader: Katherine Kiger
Indexer: Shana Milkie
Designer: Melissa Jacobson

Printed in the United States of America.
First printing August 2019.
10 9 8 7 6 5 4 3 2 1   19 20 21 22 23

Our Commitment to Green Publishing
Chelsea Green sees publishing as a tool for cultural change and ecological stewardship. We strive to align our book manufacturing practices with our editorial mission and to reduce the impact of our business enterprise in the environment. We print our books and catalogs on chlorine-free recycled paper, using vegetable-based inks whenever possible. This book may cost slightly more because it was printed on paper from responsibly managed forests, and we hope you'll agree that it's worth it. *Farming on the Wild Side* was printed on paper supplied by Versa Press that is certified by the Forest Stewardship Council.

Library of Congress Cataloging-in-Publication Data
Names: Hayden, Nancy J., author. | Hayden, John P., author.
Title: Farming on the wild side : the evolution of a regenerative organic farm and nursery /
  Nancy J. Hayden and John P. Hayden.
Description: White River Junction, Vermont : Chelsea Green Publishing, 2019.
Identifiers: LCCN 2019020272 | ISBN 9781603588287 (paperback)
  | ISBN 9781603588294 (ebook)
Subjects: LCSH: Organic farming—Vermont. | Fruit—Vermont. | Pollinators—Vermont.
Classification: LCC S605.5 .H393 2019 | DDC 631.5/8409743—dc23
LC record available at https://lccn.loc.gov/2019020272

Chelsea Green Publishing
85 North Main Street, Suite 120
White River Junction, VT 05001
(802) 295-6300
www.chelseagreen.com

To Life!

Contents

Preface	*ix*
Acknowledgments	*xi*
1. The Lay of the Land	1
2. The Early Years	17
3. Pathways to Resilience	37
4. It's All About the Soil	55
5. Adapting to Climate Change	73
6. Agroforestry in Action	95
7. Our Fruit and Nut Trees	109
8. Uncommon Berries	129
9. A Walk on the Wild Side: The Pollinator Sanctuary	151
10. Rethinking Pests, Invasive Species, and Other Paradigms	177
11. The Bees' Needs	193
12. Sharing the Farm and Farm Products	211
13. Bringing It Home	231
Appendix. Common Names to Scientific Names	*241*
Index	*247*

Preface

*F*arming on the Wild Side is the story of how and why we turned a former conventional dairy farm into a biodiversity-based regenerative organic farm. It is the story about our practices and building a relationship with the land and all its inhabitants. It describes the work that heals and restores us, the work of farming as cocreators with nature.

Society's current trajectory is leading the world to a period of increasing instability and suffering. Business as usual is clearly taking us to an evolutionary dead end. We like to think of ourselves as a pocket of resistance and an alternative success story to modern agricultural systems that degrade ecosystems and society. We know we are not alone in our desire for a regenerative, mindful approach to stewarding the land. We hope to encourage more land managers—whether farmers, gardeners, suburbanites, or urbanites with a patio or houseplants—to cultivate a renewed sense of purpose by reconnecting, rewilding, and regenerating the land and the life it supports. Right now, we can all make changes, both in ourselves and in our communities, to adapt to and help mitigate the environmental, economic, social, and spiritual crises we face.

Psychologists and counselors now use terms such as "climate anxiety" and "ecological depression" to describe specific mental health problems that are on the rise. We can relate to these feelings, especially as we reflect on tragedies in the world and the turmoil that our children and grandchildren will inherit. The fact that these scenarios are preventable is both frustrating and empowering. Maybe we can't stop runaway climate change, loss of biodiversity, and overconsumption, but we can learn to adapt to and affect these issues and ourselves with small positive acts—every day. The antidote is action in behalf of *Life*. Shout, protest, get arrested, make corporations inconsequential by not buying their crap—we are all for that. But more important, conserve or plant something! By nurturing and growing it, you and the planet will benefit from these healing actions.

Acknowledgments

A heartfelt appreciation for our friends and families, especially our parents and our children. We are grateful for all their love and support. We also want to acknowledge the hard work and positive attitudes of our employees over the years. We couldn't have done it without you. Thanks to all the visiting students who helped out with picking, mob weeding, and other often tedious jobs that became easier with dozens of helping hands. It was fun, but it wasn't always fun. Family, friends, employees, and visitors have all been an integral part of the evolution of this farm, helping to shape it into the unique place that it is. Thanks to Alisha, Sarah, and Vic for reading the manuscript and making thoughtful suggestions. We'd like to thank the folks at Chelsea Green, particularly our editors, Michael Metivier and Fern Bradley. Fern visited the farm and encouraged us to write this book. This project was supported in part by an award from the Vermont Arts Council and the National Endowment for the Arts.

FIGURE 1.1. View of the back meadow, farm buildings, and Green Mountains. The peak in the distance is that of Mount Mansfield.

CHAPTER 1

the lay of the land

It's early morning or midday or evening, any day in every season no matter the weather. It's time to walk the perimeter of the farm. Time to walk the dogs. On these daily walks, we check out what's happening with the fruit crops, the stream, the plants, and the wildlife. We notice the bumble bees on the honeyberry flowers in early May, the first milkweed flowers in June, and hundreds of green darner dragonflies massing for their fall migration at the end of August. We hear the honey bees in the goldenrod in fall and the *rawk* of the ravens in winter. We smell the delicate sweet plum blossoms in spring and taste the juicy fruit in late summer. They say the best fertilizer is a farmer's footsteps. Walking the land is the best way to understand and appreciate biodiversity too.

These walks provide a slow close-up view to the natural world and keep us in touch with plants and animals, weather, and the many happenings on the farm. A great way to start and finish the day, it's also an important step in learning about the various habitats on our land, the plants and animals that live there, and their complex relationships—the definition of "ecology." Watching our own lives unfold along with the phenology (the plants' and animals' seasonal cycles) reinforces our sense of place and connectedness with our world.

Every Landscape Has a Memory

Our eighteen-acre farm is nestled in the Lamoille River valley in the foothills of the Green Mountains in northern Vermont, just north of Mount Mansfield, the tallest mountain in the state. For a small farm, we have a wide range of soil types, from areas of almost pure sand to silts to an occasional clay lens. We have wet, spongy soils and dry areas that are susceptible to occasional drought. The variable soils and the undulating topography create a variety of niches for plants and animals. They also influence what we can grow and where. The landscape shows its past, and taking the time to learn about it has helped us better understand the land and our speck of time as its inhabitants.

Mount Mansfield was formed about four hundred million years ago, when continents collided and ocean sediments smashed and uplifted to become the Green Mountains, the range that runs north-south for the entire length of Vermont. The resulting metamorphic rock (mostly schist in our area), with its quartzite veins, is often exposed on the nearby hilltops and mountains. In our back meadow this schist bedrock underlies the topsoil, which is not very deep. In the hills behind the farm, rocky ledges jut out from forested slopes. The soil is relatively thin in these hills and is held in place by the roots of trees and shrubs.

For millions of years, erosion from wind, rain, plants, animals, and ice wore away at the uplifted rock, depositing it along the slopes and in the valleys. The last ice age started about 2.5 million years ago, with subsequent intermittent periods of warming and cooling. During some of the coldest periods, the ice might have been two miles thick in our area. As it melted during warmer periods, it left rocks and boulders in the fields that now are repurposed as the foundations of our farmhouse and outbuildings and the stone fences crisscrossing the hills behind the farm. The most recent warming period of this ice age began about fifteen thousand years ago. It's hard to imagine a hundred years ago, let alone millions. Yet the evidence is there, and the story is told in the soil-filled valleys and rounded mountains.

When it rains on the hills north of the farm, much of the water runs off in a bountiful array of seasonal waterways. A portion of the rainwater infiltrates the soil, flowing downward until it hits the bedrock. Unable to infiltrate deeper, it flows along the rock until it finds places to seep out. There are several seeps on our farm. The one in the woods, near the edge of our property, was dug out decades ago, lined with stone, and covered to form the spring box and water supply for our farmhouse. Another seep, once part of the original owner's property but now on our neighbor's property, used to flow into the barn when it was full of cows. With little buffering capacity, the thin soils and our delicious spring water reflect the acidic pH of the rain. Other seeps result in many wet, squishy places on the land.

An interesting feature of the surrounding area is that our part of the Lamoille River valley includes a gigantic "bowl" formed by the surrounding hills and mountains. About twelve thousand years ago, as the glaciers receded, a lake formed in this bowl—the ancient Lake Lamoille. Back then, the farm, the nearby villages, and surrounding lands would have been at the bottom of a giant lake! This geological event helps explain many of the features on the farm: the sandy hill on which the house stands, the silty hillocks and soils in the back meadows, and the heavy soils in other fields. All these features are results of deposits from the rivers and streams coming into this giant lake thousands of years ago and of the lake's subsequent receding into the Lamoille River.

The story of the land comprises both the geologic events still unfolding and the human influences over the years, which likewise continue. Humans came to Vermont about eleven thousand years ago, hunting caribou and other big game on a tundralike landscape and fishing in the streams and rivers. Over time, the land became forested, the caribou disappeared, and the Paleo-Indians adapted into a woodland culture. About three thousand years ago marks the beginning of the Abenaki culture. The early Abenaki people often set up camps and settlements near rivers, where they had access to fish and water. The Vermont State Archaeologist has noted that the fields and land around our farm could have been such a place. He ruled that before the neighbor's meadow could even be considered for development, an extensive (and expensive) archaeological investigation would be required. We were happy with his recommendation, although our neighbor, who was hoping to develop the site, wasn't.

The first colonizers of European ancestry arrived in this area in 1793. The farm was settled a few years later, around 1809. The newcomers of the 1800s spent the next fifty years or so cutting down the forests that covered Vermont. They set up sawmills for wood and wood products. They cleared land for planting and creating pasture for sheep, mules, horses, and cows. The old stone fences from their farms still crisscross the hills behind our farm and mark our property boundary on our western side. By the late 1800s, only about 20 percent of Vermont was forested; the rest had been cut down. Today, about 80 percent of Vermont is forested again. These new-growth forests continue to be logged periodically, but usually selectively. Most of the trees in the neighboring woods behind our farm are relatively young, as evidenced by their smaller sizes.

The homestead that started on our land in the early 1800s supported various livestock operations over the years and eventually a dairy that lasted until the 1980s. Now it is a perennial fruit farm. We wonder what it will look like in another hundred years.

Every Farmer Brings Their Perspective

Immersing ourselves in nature has been our basic need and principal goal ever since we were children. On the farm, we spend the better part of our day, for eight months of the year, working and being outside in all kinds of weather. We note what's in bloom, what insects and other wildlife are doing, and which fruits are ripening. We pay attention to the world, and it has never failed to fascinate, teach, and entertain us while at the same time nurture us.

Our love of being and working in nature has only grown over the years. It was what prompted both of us, coming from different parts of New York State, to study biology and ecology at the SUNY College of Environmental Science and Forestry in Syracuse, New York. Our program of study not only brought us together in our junior year of college but also encouraged and supported our love of the natural world. It helped us better understand biology, the study of life, and ecology, which focuses on the relationships between organisms and their environment. Ecology emphasizes that we are all connected to this wonderful planet Earth, whether we're a tiny insect or a human being. We all *matter* too. This holistic view instilled in us the importance of ecology as a lens for thinking about and appreciating the world.

After college, we continued our exploration of the world and its many inhabitants by joining the Peace Corps. The opportunity to learn about the world, about people and cultures, animals and distant lands, as well as to work with others, motivated and inspired us to join. Nancy partnered with Kenyan fisheries' agents to work with subsistence farmers and women's groups to promote the building of family-sized fishponds. On the other side of the continent, in Mali, John spent his two years working to help small farmers with the management of millet pests and market gardens. The results of our service were broadened worldviews, personal reflection during solitude and loneliness, cross-cultural friendships and interactions, travel adventures, an awareness of our white privilege, and connections with new landscapes and ecosystems. We grew emotionally, intellectually, and spiritually.

We rejoined forces after our Peace Corps service, got married, and started graduate school at Michigan State University (MSU). John studied in the Department of Entomology and conducted research on the ecology of apple pests. He also got an insider's view of the pesticide industry and large-scale commercial fruit farming. Nancy furthered her studies in environmental engineering and learned about the fate, movement, and toxicities of pesticides and other chemicals in the environment. Our combined education and graduate experiences greatly influenced our organic, no-pesticide philosophy.

While Nancy was pursuing her doctoral studies at MSU, John spent a few more years in entomology research and later moved into agricultural extension. He worked with farmers, implementing integrated pest management (IPM) and best management practices for protecting streams and rivers from agricultural runoff. He oversaw a three-thousand-acre crop management association of corn and soybean growers and was able to reduce their use of agrochemicals; however, he grew tired of preaching about what farmers should do to an often-skeptical choir. The chemical and fertilizer companies, with their marketing and fearmongering, were tough to compete with—especially when farmers felt as though they were living on the edge and needed to be risk averse. Rather than continuing to tell other farmers what they should do, John had his heart set on starting his own farm, where he could explore different techniques for growing food and test his agroecological ideas.

When Nancy received an offer to become the new environmental engineering faculty member—and first woman faculty member—in the Department of Civil Engineering at the University of Vermont (UVM) in 1991, we packed up the family and moved to the Green Mountain State. We bought our farm the following spring.

Our desire was to create a viable alternative to an agricultural system we considered broken, one that we believed was leading society on a problematic trajectory of increased soil and water degradation, pollution, and ecosystem destruction. This food system brings society cheap food, yes, but also diet-related diseases, poor nutrition, obesity, and waste. We had a different idea, one of growing healthy food for our family. We also appreciated the physical work that kept us present in the body, connected us and our family to nature, and helped our mental and emotional well-being. Today people talk about the importance of "forest bathing"—getting out into the natural world for their own health and peace of mind. We have been lucky in that regard. Farming has created a means to incorporate this therapy as a lifestyle.

Our farm is a big deal to us, even if it is but one small part of the interconnected regional landscape we call home. Yet how we attend to the small daily aspects of our lives—our work, nourishing food, daily exercise, and an attitude of wonder and curiosity about the natural world (ecological thinking)—helps with the big things too. Big things such as climate change mitigation, food sovereignty, and social justice.

Every Farm Has a Story

We welcome hundreds of visitors and customers to our farm each year. Their first glimpse as they enter the back driveway often elicits exclamations of surprise and wonder, especially if they are newcomers. Back in 2009, we turned the once extensive front and side lawns next to the driveway into an apple orchard, flanked with rows of lingonberry on one end and rows of gooseberry, red currant, and honeyberries on the other. A diverse collection of native trees and shrubs grows along the stream bank and in a demonstration garden we call "the pepinyè garden." Most people don't expect a hidden enclave of biodiversity just off the busy road. It's all a far cry from the manicured monoculture lawns, fields, and ditches that were here when we bought the farm twenty-seven years ago. Observing and then mimicking nature in our farming practices has been an important part of our agricultural approach.

It's also hard for visitors to miss the historic post and beam dairy barn and its companion silos, which command immediate attention on the drive in. After all, a big red barn is what both newcomers to Vermont and old-timers expect to see on a farm. Built around 1900, with a cathedral hayloft and a steep-pitched slate roof, ours is a reminder of the farm's former dairying days. We appreciate the craftmanship and beauty of the old barn. We marvel at the trunks of the giant hemlock trees that became the hayloft's hand-hewn posts and beams. Yet it is also a reminder of the more recent confined cow operation and of unsustainable agricultural practices, as well as an era gone by. In our early years, we used parts of the barn for hay storage and for livestock, although nowhere near its capacity. Its vast size has encouraged the worst of our hoarding instincts over the years and is loaded with scavenged and saved supplies, equipment, and other materials, but even now, it is nowhere near full. It still has a lot of life left to it, though, so we keep it maintained. It's too big (and too beautiful) to fail. The looming concrete silos behind the barn are another matter. They were built by the previous owners in the 1970s, a few years before they took advantage of the 1980s dairy buyout and sold their herd. It's fitting that these great gray monuments now house a growing mound of guano from nesting pigeons. Future fertility for the land? We hope so.

Our focus on biodiversity, ecology, and wildness as integral components of our farm has been interpreted by some as unkempt and messy. We say society needs to rethink "pretty," shed the cultural conditioning, and celebrate the wild, especially in the face of current environmental problems. Sweeping grass lawns and monoculture fields have led to environmental degradation and loss of food and habitat for pollinators, other insects, and birds. This monoculture mentality creates a dependency on pesticides, which has devastated ecosystems and life around the planet. It is outdated thinking.

Instead, we're focusing on a regenerative food production system that keeps us healthy by providing nutritious organic food and meaningful work. It also heals the land by improving soil and water quality, increasing biodiversity, and sequestering carbon from the atmosphere to help mitigate climate change. Increasing biological inputs, enhancing ecological restoration, and using no-till production, along with a policy of zero pesticides or chemical fertilizers (even organically approved), go further than most organic approaches. Besides improving the environment, organic farming can be economically, socially, and spiritually rewarding as well.

We started out in grass-fed livestock and organic vegetables. About fifteen years ago, motivated by concerns about negative effects of tillage on soil health, competition for markets, and resilience in the face of climate change, we began transitioning our draft horse–powered integrated mixed-vegetable and livestock organic farm to perennial fruit production. Hurricane Irene in 2011 was our wake-up call to the reality of climate change and hastened our changeover. We planted a vegetable field that flooded during Irene with elderberry and aronia the following year. These perennial fruit shrubs don't mind an occasional flood.

Irene also prompted us to start a retail nursery of fruit trees, berry bushes, and conservation plants to help bolster our economic resilience and encourage others to grow their own food, decrease the size of their lawns, and increase carbon sequestration and biodiversity.

FIGURE 1.2. Entering the back driveway in 1992 shows a neat-as-a-pin environment with a cleaned-out farm ditch and low biodiversity—a biological desert.

FIGURE 1.3. Entering the back driveway at The Farm Between shows a farmscape of fruit trees and berry bushes on the left and a diversity of native plants protecting our seasonal stream on the right. Biodiversity is multifunctional beauty.

the lay of the land

Our nursery sales area is located between the farmhouse and the barn. It's filled with a variety of potted trees and shrubs, propagation beds for starting much of our nursery stock, and perennial demonstration gardens. Rainwater collection tanks and rain barrels are strategically located to collect rainwater from our roofs and are used for watering nursery plants. Quite a different look from the swing set, trampoline, and badminton court of earlier child-rearing days.

A portion of the barnyard on the west side of the barn is fenced off for our retired draft horse mare to use in the winter to stretch out, roll in the snow, and soak up sunny days. Farther back marks our composting area, with several piles at different stages of the process. Manure from our horse (horses and other animals in the old days), manure from the neighbor's horses down the road, weeds, household food scraps, leaves from local community members who for some reason don't seem to mind exporting organic matter from their yards, and shredded tax returns from a nearby accountant all undergo the turning and heating of thermophilic composting. The sweet-sour scent of the living compost can often be detected in the air during the initial stages or after a turning. On a cool morning, water vapor rises from the living pile, a by-product of microbial respiration, like the water vapor exhaled in our breaths. Over four to six months, all this material metabolizes into rich organic compost that we use in our nursery and for our fruit plantings. Composting is a fundamental component of our soil health practices.

On one edge of the barnyard, a perennial-planted hügelkultur (mound culture) blooms in the summer, providing nectar and pollen resources for important pollinating insects. Our open design creates nesting and overwintering places for bees, as well as habitat for beetles, earwigs, centipedes, and snakes. These mound cultures, which we affectionately call "bumblekulturs," are mostly made up of prunings from our fruit trees and berry bushes. Annual pruning keeps the plants healthy and gives us enjoyable work in the late winter and early spring. Once the pruning piles get big enough, they're covered with soil or compost and eventually planted with cover crops, flowers, zucchini, or whatever else we feel like planting. At least half a dozen of these living mounds, at different stages of the hügelkultur process, adorn the farm.

Beyond the nursery and barnyard are the various fields of perennial fruit. In field 2, rows of rhubarb with their oversized leaves are the first plants to show signs of life in early spring. Crinkled convoluted reddish-green leaves just barely emerging from the soil look like a terrestrial version of tropical coral. Toward the end of May and early June,

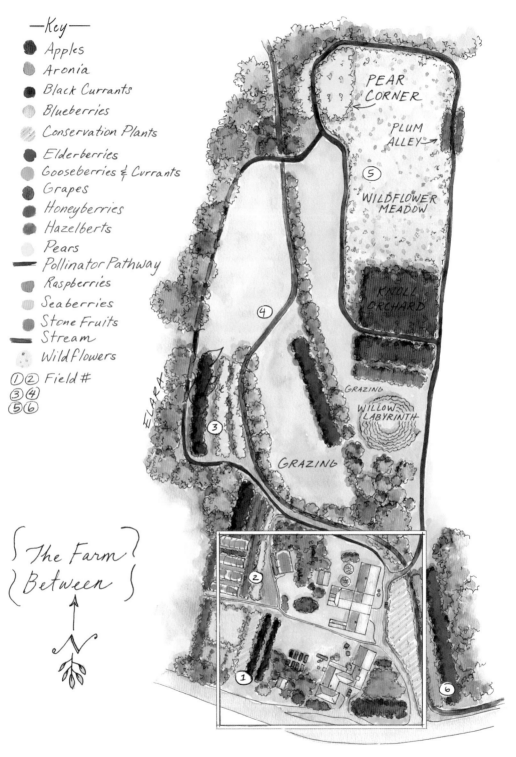

FIGURE 1.4. A map of The Farm Between shows the nursery area, various fruit plantings, buildings, and other features of the farm today. *Illustration by Elara Tanguy.*

the lay of the land

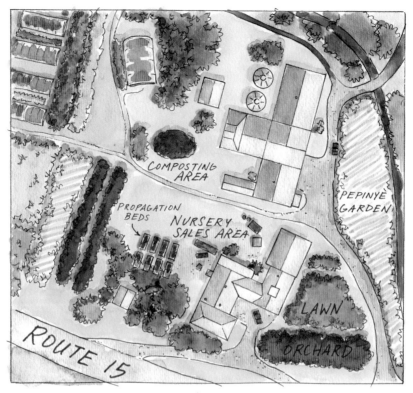

detail

we harvest hundreds of pounds of the reddish-green stalks for wholesale markets and our own use in rhubarb syrups and pies. After harvest, we'll weed it, add compost, and mulch it so it grows well for next year's harvest. This is pretty much the routine for many of our plantings. Only the timing varies, which depends on the harvest date of the crop. Late-harvested crops such as apples get composted early in the spring.

New blueberry rows in field 2 and older ones in field 1 are among the slower bushes to leaf out in spring, but they're worth waiting for. White bell-like flowers buzzing with bumble bees in late spring, sweet purple berries ripening in July, and a background of reddish-orange leaves in the fall give us plenty to marvel at.

Walking by the first of five unheated hoophouses in early September means a quick stop to pick a handful of juicy fall-bearing raspberries. We call them our "anticancer meds" because of their high concentrations of antioxidants and other anticancer properties. Picking the trellised raspberries is easier on the lower back than picking most of the other berries because we don't have to bend so much. We gradually

FIGURE 1.5. A row of beach plums blooms in spring. Hoophouses 1 and 2 are in the background, with mulched young blueberries and a rhubarb patch in the foreground.

added the hoophouses over the past dozen years as a resilience measure to help us grow quality tomatoes, peppers, and fall-bearing raspberries in the cool, often-wet growing season of northern Vermont.

For the past several years, we've been repurposing the tomato houses with fruit trees. In one hoophouse, we have dwarf high-density apple trees on trellises. In another, cherry and peach trees, and plum and apricot trees mixed into the raspberry houses. The stone fruit are still a new experiment, but the apple trees have been producing well for a couple of years now, providing us with blemish-free no-spray organic apples. The hoophouses allow us to grow high-quality organic fruit by giving us the means to control the effects of frosts, rain that causes fungal pathogens, birds and other pests.

Behind the hoophouses, a line of twelve-foot-tall (3.7 m) hazelbert nut trees forms a bright green hedge in summer and a rusty gold one in fall. In the microclimates created between the hoophouses, we've tucked honeyberry bushes, perennial flowers and grasses, and more blueberries. Other fruit crops, such as elderberry, aronia, and gooseberries, fill various fields and niches around the farm based on the soil and moisture conditions, microclimates, and our ability to access them.

The first berry bushes we planted for production were the black currants in field 1. We then added more rows in field 2 and field 3. Most of the black currant crop goes to local wineries and breweries, but we also use them for our own value-added products (fruit syrups, jam, and the like). Occasionally, we sell a few pounds fresh to other black currant lovers. We recently planted more of these deer- and bird-resistant black currant bushes in our perennial polyculture orchards within the fourteen-acre (5.7 ha) parcel we call "the pollinator sanctuary." In our early years of farming, we called these fourteen acres "the back pasture" and grazed sheep, heifers, turkeys, and horses there. We hayed part of it, too, but that seems so long ago. Since 2008, we've been rejuvenating it and turning it into a pollinator and wildlife sanctuary. The work to reclaim wet, reed canary grass–dominated areas with willows, silver maples, and other native trees and shrubs is still in progress, but we're getting there. We stopped grazing and haying other parts of the pasture, and they have reverted to wildflower meadows of milkweed, goldenrod, and aster. We created vernal ponds and pools for frogs and dragonflies. We planted a small one-acre heritage apple orchard on a silty knoll (Knoll Orchard) and a pear orchard in the back corner (Pear Corner). Black currants, bayberry, mountain ash, and other conservation trees and shrubs are mixed within the tree rows. We added perennial flowers such as penstemon, *Agastache*, and sneezeweed between the rows for the pollinators as well.

Recently, we've been experimenting with alley cropping, planting perennial vegetables and annual hemp in the alleys between the trees. We leave other alleys unmowed for much of the summer to allow the milkweed to grow. We've harvested it and sold seeds and floss for a few years now. We started a willow labyrinth and finished planting it in 2018. We've also sold materials from many of the native shrubs we've planted to florists and wreath-makers. So many rewilding endeavors have kept us invigorated and the wildness in our hearts alive.

The edges of fields, riparian areas around the stream, and other hard-to-get-at spaces on the farm have also reverted to wildness, which we've often helped along by planting a few of our favorite native trees and shrubs appropriate for that location, such as black locust, swamp white oak, and buttonbush. These wild places provide biological diversity, habitat, and food for insects, birds, and the other important players in our ecological pest management approach. They provide usable biomaterials for mulch, woodchips, and other products. They can even help with carbon sequestration. Imagine planting the tens of millions of acres of little-used grass lawns in the United States with woody shrubs and trees. That would mean a lot of carbon taken out of the air.

The story of our farm today is one of multiple hedgerows, habitat corridors, and beneficial insect and other wildlife habitat within a perennial polyculture fruit-oriented farming system. Like us, it is a dynamic system. It's also a practical application of our philosophies, passion, ecological understanding, spiritual awareness, the exuberance of our youth, and now, the tempered wisdom of cumulative experiences and aging. We're on the continuum between the old way of farming moving toward a new regenerative paradigm. Have we arrived? Hell no! There are always new things to learn and new ways to grow.

FIGURE 2.1. ElectroNet fencing was used for our sheep management intensive grazing system.

CHAPTER 2

the early years

Stepping outside onto the back porch of the old farmhouse and looking over the farm, there's a moment, or perhaps several, of pure perception. We breathe it all in: the coolness of the morning breeze, the rustling of leaves, doves cooing, and the sweet scent of lilacs. It's a bit like that feeling you had as a kid, stepping outside on the first day of vacation, smiling at the summer and all its possibilities.

We see the fruit fields and hoophouses, the distant wooded hills, the historic, polygon-shaped corncrib. How many times have we stood on that back porch, softening our eyes as we gazed at the farmscape or focused our hearing on spring peepers or a great horned owl calling out from across the river, feeling part of the life around us and exhilaration at the life within us? Thousands? And it never gets old.

Our view is always changing, through the seasons and years and the growing and dying of things. We're changing too. Our interests and priorities, our aging bodies, and outside factors influence our decisions and often move us in different directions. One of life's important lessons is that nothing ever stays the same. Impermanence. That can be a blessing in tough times.

Farming, like life, deviates from predicted paths. Sometimes it happens through a random event: a flood, or someone giving away free sheep. Other times, lifelong interests final bubble to the surface, such as getting draft horses or planting apple orchards. Looking back, we might try to describe our farming adventures as a well-laid-out progression, but in many ways, it was more of a dance. Alan Watts, a twentieth-century philosopher and writer, likened life to a dance or music, rather than a journey where the end, the "getting there," is the important point. We like that analogy. For us, farming has been at times a fast and furious dance to rock and roll, a slow waltz, and everything in between. We can try to conduct the music, but usually it conducts itself, and then we decide how best to dance to it.

A Holistic Approach

Starting the farm that first summer was both exciting and a bit overwhelming. There were so many possibilities. We had several important priorities. We wanted to feed our family high-quality organic food: vegetables, fruits, eggs, and meat. These products were difficult to find in the early 1990s. They weren't available in supermarkets. Co-ops were small and not too common, and farmers markets and community-supported agriculture (CSA) were still in their infancy. If

FIGURE 2.2. Occasionally, a lamb was rejected by her mother and had to be bottle-fed. Connor cuddles a friendly bottle-fed lamb while Nolan communes with a laying hen.

we wanted organic, we were best off growing it ourselves. Not only did we not want to feed our children food laden with potential pesticide residues, but we also didn't want to contribute to the environmental issues associated with them. For millennia before the twentieth century, people grew high-quality food without pesticides. We knew we could, too, and we'd be able to sell it in our community.

Other important goals were to treat our animals humanely and manage the land with future generations in mind, leaving it better than it had been when we bought it. We also wanted the farming operations to generate income. It wasn't just a lifestyle for us; we wanted it to be a livelihood too. Our love of being around animals and the desire to share that passion with our children was another priority. These nonhuman partners helped connect all of us to the farm, to food, to the seasons, and to life's cycles of birth, death, and regeneration.

Early on, we took inspiration from a weeklong holistic resource management (HRM) workshop that explored a goal- and mission-oriented systems approach to farming. To us, a systems approach means that our farm and the elements it comprises do not exist in a vacuum. The farm and all its parts (including us!) are linked to one another and to the surrounding ecosystem. This systems-thinking approach places not only ourselves, but also our family and our community within that ecosystem. It means that the choices we make, including our farming strategies, have long-term and outwardly rippling effects beyond the here and now. This tied directly into our ecological thinking, our Peace Corps experiences, and our graduate studies, which all emphasized that systems (even our global ecosystem) are integrated and connected.

We were also inspired by many small-farming pioneers, including Booker T. Whatley, Eliot Coleman, and Joel Salatin (the latter of whom John visited in the spring of 1992 to see his pastured-poultry chicken tractors, before Joel became a much sought-after rock star), and learned a lot by trial and error and keen observation of the animals and plants in our care. Over the years, we raised sheep (for wool and meat), pigs (for meat and feeder piglets), chickens (for eggs and meat), turkeys, rabbits, a family cow, and a full range of vegetable crops. We sold our meat through our meat CSA, and our produce, eggs, wool, and other items at our farm stand. Our farm back then was not so different from many small-scale diversified farms in operation across New England today.

Our farming ventures have not always taken a well-laid linear path, as we have long farmed by the ecological tenet, "Diversity equals stability." From the start, we have valued diversity in our vegetable crops, animals, farm products, markets, and farming approaches, as

DEVELOPMENT OF OUR DIFFERENT PRACTICES AND EVENTS OVER TIME

Early years (1992–2001)

Focus on regenerating soils and pastures
Integration of animals with vegetable production
Meat CSA (chicken, turkey, lamb, pork)
Farm stand and on-farm agritourism for six years
Farmers market at Smugglers' Notch Resort
1995 flood
Field trips from local elementary and preschools
Nancy's promotion to associate professor at UVM

Middle years (2002–2011)

Dedication to resilience issues, pollinators, and biodiversity
Draft horse power
Uncommon fruit plantings
Expansion of perennial and annual crop production, cutback in meat production
John's position as lecturer in the Plant and Soil Science Department at UVM
Veggie CSA on the farm for five seasons, along with different farmers markets
Five hoophouse additions over this time period
Development and marketing of fruit value-added products
Establishment of Front Lawn Orchard and beginnings of Knoll Orchard
Foundation of the Seeds of Self Reliance nonprofit
Kids' departure from home, the start of their independent lives
Field trips for special-needs children from local schools and summer camps (middle school ages)
2011 floods

2012–present

Commitment to perennial polyculture, biodiversity, and pollinator protection
Tractor purchase and retirement of horses (for most purposes)
Nancy's retirement from UVM
More fruit products and cider development
Burlington Farmers Market
Nursery on the farm
Education around fruit and conservation plants
On-farm workshops, field trips, and other educational opportunities for high school students, college students, and adults

well as in our family experiences. If it seemed that we were all over the map, everything was part of the farm flow, connected even as new projects were unfolding. Our enterprises on the farm often shifted as a result of changing dynamics within our family, our community, or ourselves. Yet everything we've done has always been about creating a positive relationship to the land and taking small actions that accumulate over time toward enhancing that relationship. The lifelong learning accrued through our different practices (and studies) has been priceless.

Integrating Livestock into an Evolving Farming System

One of the essential components of HRM that worked for us in the early years was management intensive grazing. It's worthy of a little discussion here because it's still important on the farm today. Management intensive grazing is a method that allows you to constantly improve your pastures by managing the number of animals per unit of ground (stocking density), which is based on the growth rate of the pasture. Because plant growth rate changes over the duration of the season, HRM practitioners change the paddock size accordingly to ensure the animals are eating everything in a short period of time. The main idea is that if you leave the grazing livestock in a larger area with more grass and forage than they can consume, they are going to eat their favorite, most palatable plants first and leave the others. This puts natural selection to work in favor of the *less* desirable plants. You can run down pasture quality easily by allowing your livestock to be choosy. On the other hand, by using management intensive grazing and having the animals eat everything uniformly, you can turn a low-quality pasture back into a lush and nutritious field for your animals.

Grazing is a magical system. Animals gather their own feed and spread their own manure on the pasture. They get fresh air and can eat what they want (even self-medicating with certain plants). The excitement of the animals when they get let out onto spring pasture after a winter in barnyards or even as they move to a fresh slice of pasture is contagious. They exude health and happiness. It's hard to believe that so many of society's current animal-raising approaches confine animals 24-7 to feedlots or pole barns.

After we bought the farm, we allowed a local farmer to keep haying for several years in exchange for winter feed hay for our increasing livestock herds. We put a stop to his burning of the previous year's dried grass residue, though. It turned out that was his usual approach in the spring to create potash. We wanted the organic matter to go back into the soil, not the atmosphere. While the soil structure on the farm was still good in terms of tilth and drainage, the fertility was low. Low fertility and soil organic matter (less than 3 percent) were confirmed by samples sent to the UVM soil-testing lab. The pH was also low, meaning the soil was acidic and would not be optimal for the pasture growth we wanted. So we hired a local dairy farmer to spread limestone (calcium carbonate) to increase the soil pH. That was the easy part.

The pasture was also overrun with reed canary grass, a relatively unpalatable and not very nutritious grass for animals. To be fair, reed canary grass does have its benefits. It grows quickly and in wet areas. A few northern European countries are looking into using it as a biofuel. Other applications include using it to soak up nutrients in wastewater treatment applications or as an introduced plant after clear-cut logging operations. However, on the farm, it was aggressive and outcompeting other grasses and forbs, which have a greater nutritional value for grazing animals, creating what was in our view an unproductive biological desert. Within the first year of management intensive grazing with our small flock of sheep, however, we saw clovers, other grass species, and herbaceous forbs such as dandelion, Saint-John's-wort, and plantain gaining a toehold over the reed canary grass in the drier areas.

Within our first two years, we also started raising chickens (for meat) in portable chicken tractors on our fertility-challenged land. The birds followed the sheep and ate up fly larvae in the manure, scratched and spread that manure around, all while adding their own manure to the biofilm of soil we were trying to enrich. The chicken tractors that we moved twice a day were also a much better living environment for the chickens than that of conventional chicken operations, which pack chickens into tiny spaces that consequently require antibiotics for disease control and the cauterization of their top beaks so they don't peck each other. We wanted our chickens to get fresh air, fresh pasture, and fresh water, as well as peck for critters in the soils and grass.

We adjusted Salatin's design after our first season by making a hoophouse-style tractor. This made it easier to get inside the tractor and stand up to work, while increasing air ventilation and cooling for the chickens. We used this style of chicken tractor for more than a dozen years, raising thousands of chickens. After the first season, we also switched our cockerels (males) from disease-prone and sluggish

FIGURE 2.3. Hoophouse-style (10′ × 14′ / 3 × 4.3 m) chicken tractor filled with Kosher King cockerels. Hoops made from 16′ (4.8 m) schedule 40 1″ (2.5 cm) PVC pipe were held in place with conduit clamps. Chicken wire was attached on the sides, with rain and sun protection above. Feeders and gravity-fed waterers hung from PVC ridge purlin.

Cornish/White Rock crosses, the industry standard, to Kosher Kings, a Cornish/Barred Rock cross, which were also much more active and better foragers. This rugged breed ate less grain but took longer to mature (ten to twelve weeks). We considered that a fair trade-off. They also had a fierce personality and chased down bugs and weeds like little feathered velociraptors.

An important concept we learned by studying and practicing various grazing strategies was that as plants grow aboveground, their roots grow correspondingly down into the soil. This is beneficial because the plants can access more water and nutrients deeper into the soil, which in turn extends the whole root ecosystem biology deeper into the soil. When the plants are cut or eaten down, they slough off those deep roots, thus providing organic matter and food sources for soil organisms to consume in an increasingly deeper biologically active zone. This is particularly important in our orchards, and is a good reason why we let the grass grow tall around our fruit trees, cutting it only once or twice a season as a mulch.

FIGURE 2.4. We also raised turkeys on pasture with an open hoophouse design that was for rain shelter and keeping the grain dry. The ElectroNet fence kept predators out except for owls, which were one deciding factor in the ending of our turkey operation.

Laying hens were the first animals to join the farm and the last to leave. It's likely they'll be making a comeback soon. It's no surprise to us that they have been living in partnership with humans for upwards of eight thousand years. They're good foragers for insects, seeds, and plants, and turn food scraps into eggs. Our free-range hens certainly liked to explore the pasture, scratching the manure of whatever animal they followed in the leader-follower intensive grazing scheme, breaking up cow patties or horse apples to search for fly larvae, or scavenging for the earthworms that enjoy living in or under the manure.

During the growing season, our hens typically grazed in a fenced area and returned to the wheeled eggmobile in the center of their paddock in the evening or when they laid their eggs in the nest boxes. Even newcomers quickly learned how to run to a new paddock when the fence was opened. The eggmobile was easily rolled to the new spot. We raised mostly Rhode Island Reds or similar breeds, but a few Araucanas that lay green eggs and a couple beautiful Cuckoo Marans that lay chocolate-brown eggs joined the group at different times, picked up from another farmer's hen downsizing.

FIGURE 2.5. Laying hens and their eggmobile take a run through the young Front Lawn Orchard to help with insect pest management, reduce grain costs, and make beautiful eggs from food scraps.

After we figured out systems for raising sheep and poultry, we started buying eight-week-old weanling pigs for us to raise for our own table, and to sell in our meat CSAs. In just six months, eating a combination of organic grain, pasture, weeds and vegetable leftovers, they grew from little pink "cute as a bug" muscle packs full of energy and squeals into NFL-linebacker-sized muscle packs full of snorts and grunts and ideas of their own. One thing that stayed the same was they loved getting their bellies scratched.

As with all the animals on the farm, our first order of business was to train them to the grain bucket and to an electric fence. After we shook the bucket of grain once and gave them a taste, the pigs were trained to come to the sound. The electric fence took a time or two. Pigs don't usually jump much, so two strands of polywire at six inches and eighteen inches (15 and 46 cm) from the ground kept them in and were easy to move in a rotation. Pigs occasionally tested the electric fence—as evidenced by their sporadic squeals when getting shocked—so we kept it hot. The high voltage but low amperage made it safe, but a jolt nonetheless. (We know from lots of experience.) When on sod, the pigs used their snouts as shovels to throw sod and soil on the fence and ground it out. An accident? We never thought so.

After a few years of buying young feeder pigs for raising, we had the opportunity to buy Gloucester Old Spot–cross sows, a heritage breed of white pigs with black spots. We bought four, borrowed a stud boar named Dudley from a farmer friend, and were off and running in the piglet-raising business. We sold feeder pigs to other farmers and kept a few to raise for ourselves and our customers.

A farm in Vermont, especially one that was once a dairy farm, wouldn't be complete without a family cow. Honey was our first, then Betty. We really enjoyed our cows over the years. They're calming animals. The conventional method is to take the calves away for good after they get their first meals of nutrient- and probiotic-rich colostrum, and then milk the cow twice per day. Our more humane method for raising them was to take only the morning milk. That was more than enough for family and friends. After we hand-milked Honey in the morning, we would put her back together with her calf. Hope was her first. Hope nursed the leftovers and made sure that Honey was all milked out. There's less chance of mastitis that way, and we didn't worry about getting every drop of milk from Honey. Hope could also nurse all day (including the evening milking) until our final evening chores were done. Then we'd separate the two of them into two adjacent pens. Hope could see and smell and hear and even touch her mom but couldn't quite nurse. The next day, we'd milk Honey and start the routine they became comfortable with again.

There are so many benefits to this way of having a dairy cow. First off, mother and calf remain together in those early weeks, which keeps them both so much more content. The calves grow bigger and stronger than if they were raised on a milk bucket with nipples, which in both organic and conventional dairies typically starts hours after they're born. When we did custom grazing for an organic dairyman, we could easily compare his six-month-old heifer calves to Harvey (Honey's second calf with us). His heifers were about half Harvey's size, and it wasn't only because Harvey was a male. When a respiratory ailment worked its way through the same group, Harvey didn't even get slightly sick. He was so strong and healthy. Unfortunately, many of the young heifers raised without their mothers weren't so lucky.

FIGURE 2.6. A sow and her brood "hogging down" a rye-vetch cover crop.

Slaughtering Animals

While there were a lot of benefits to raising animals, there were also some challenges. For one thing, animals got out even though we were careful about fencing. That's why all our animals were initially trained to come running when they heard grain rattling around in a bucket. Since we live close to a busy state road, animal escapes could be pretty nerve-racking. Four-hundred-pound sows and an even larger boar licking salt on the side of the road or horses running down the middle of it might have spelled disaster if we didn't intervene quickly. We only ever lost a black cat to the road, though, and luckily our escapees never caused anyone else to be hurt. Bucket-train your animals: an important safety tip.

Raising a lot of animals also made it tough for both of us to vacation at the same time. Animals need to be cared for daily, fed, watered, moved during the growing season when grazing, and regularly checked on. We've been lucky over the years to find farm-sitters for special trips, but at other times, one of us would have to stay to care for our animals.

Slaughtering is also tough. We never treated our animals as nonsentient, instinctual beings, as some people may. We cared for them, recognized them as having consciousness and personality, and tried always to remember their ultimate sacrifice. Because of that outlook, there was a bit of sadness and resignation for us at slaughtering time. It was hard to find a slaughtering facility that we felt treated the animals in a humane way before slaughtering them. Transport is stressful for animals. We opted for home slaughter for chickens, turkeys, pigs, and many of our sheep to maintain control of the whole process. On-farm slaughter regulations are continually changing, though, which presents challenges for small farmers who are committed to humane, low-stress slaughter.

Slaughtering our pigs was especially tough. We couldn't help but be friendly with them. They loved attention and even let the kids ride them when they grew bigger. They came running for treats or to see what else we might have for them. We never had so many that we lost sight of them as individuals that we got to know. We didn't name them, but still we knew each one and liked them a lot. It was hard to say goodbye.

We opted for home slaughter for the pigs, which meant custom butchers coming to the farm to kill, skin, and eviscerate them, and take the carcasses back to their place for custom cutting based on our customers' orders. We never looked forward to slaughter day, but we would try to make it as easy on our pigs as possible. Pouring milk into a trough so they would be happily slopping it up when the .22 bullet entered their brain was our best strategy. The butcher then jumped into the pen and cut their jugular vein to bleed them out. Our experiences in getting to know our animals and also being present at their slaughter has led us to believe that people should similarly know what it means to eat meat.

We discovered an interesting way to market our lambs and sheep toward the end of our sheep-raising days. Vermont is small, so when a group of immigrants who practiced a particular tradition of on-farm slaughter (halal) in their home country found out about our flock, they contacted us directly to see if they could do the same thing here. We were able to bridge the language gap using sign language and a shared, innate understanding to make the slaughter as calm and quick as possible for the animals. These same immigrants came to the farm on several occasions to catch and kill the animals, taking every part of them for further use except the hooves. This seemed a much better end of life for the animals than being loaded into a truck and sent to a USDA-inspected slaughterhouse where they would be stressed and handled roughly. This cross-cultural arrangement happened years ago. We are pretty sure it is not legal anymore. Maybe it never was. Too bad.

While we were small-scale livestock farmers for many years, we weren't necessarily hard-core meat eaters. It's not a bad idea for Americans to scale back on meat eating, since our average consumption is about 40 percent higher than USDA *2015–2020 Dietary Guidelines* recommend. As biologists, though, we recognize that humans are omnivores and have evolved and survived by eating meat for hundreds of thousands of years. We love our vegan and vegetarian friends and respect their choices, but we think of grass-fed and pastured meat as an important part of our diet and farm ecosystem.

With that said, the current prevailing system of animal agriculture using confined animal feeding operations (CAFOs), or factory farms, is an abomination from an ecological and ethical standpoint. Raising animals in a manner that relies on the overuse of antibiotics to counter the unnatural living conditions and monoculture feedstock diet of mostly corn and soy while exploiting human workers and degrading the environment through energy use and pollution makes society's high-animal-protein diet a downright disgrace.

We tried to do better. We tried to teach our children to appreciate their food by understanding the work and sacrifices inherent to the process. It's easy to take the meat on the table for granted when you've never known the animals that produced it. It's a lot harder when you've scratched their ears and watched them romping in the field or grazing on grass only a short time before.

Changing our meat-eating habits to consuming primarily humanely raised grass-fed animals has made it difficult to go out to enjoy conventional beef, pork, and chicken entrées in restaurants. In Vermont, we're lucky in that we have a few restaurants nearby that include locally raised meats on their menus, but it's not always the case, so we eat at home a lot or opt for the vegetarian dishes when we go out. What we buy and what we eat are political acts. When people buy our products, they are helping us farm regeneratively. When we buy something, we are supporting the business that's behind the product. Voting with our wallets. It's not always easy or inexpensive to support local, organic, regenerative, and fair trade practices. It might depend a lot on where you are and where you live, too, but it's important that all of us keep trying to make positive changes to our food system.

Draft Horse Power for Self-Sufficiency

After working with Clydesdales for one summer at Busch Gardens during college, John became interested in draft horses and set out to learn as much as he could about horse power. In Michigan, he attended workshops on draft animals run by Tillers International and worked with older mentors who were still using horses for timber harvesting or farming. He read *Small Farmer's Journal* from cover to cover every month. In Vermont, he joined the Green Mountain Draft Horse Association and attended a variety of seminars and workshops, including a three-day intensive with Jay and Janet Bailey at Fair Winds Farm near Brattleboro. Visiting their horse-powered farm and witnessing their relationships with the horses was cathartic and inspirational.

After twenty years of learning and practicing, John finally felt ready to have his own team. That happened about ten years after we started farming. This was also the time when we wanted to expand our vegetable and small-fruit operations and had realized the limits of human-powered labor. We didn't have or want a tractor. We felt that it was a good time to realize the draft horse dream.

Incorporating draft horses into the farm was also a good fit for our ecological small-scale farming mind-set. The opportunity to learn about and bond through teamwork and care with these majestic, powerful animals, slowing down the work by grooming and harnessing before and after, raising a foal, and all the wonderful manure meant more to us then than the efficiency a tractor could have provided. Horses not only improved our quality of life but were incorporated into our grazing rotation and resulted in less soil compaction while plowing, discing, cultivating, and manure spreading. Draft horse power was an appropriate technology for our scale of farming and goals. Affordable and sustainable with discarded repairable equipment available at auctions and in hedgerows, it fit our budget, interests, and desire to partner with these complex and willing beings. In addition to farming, we skidded firewood with them, gave family and friends sleigh and wagon rides, rode them, and loved them. Nora and Nellie, the team of mares we found in Canada, were an awesome addition to our farm.

When tractors replaced horse power a hundred years ago, efficiency, as measured in crop yield per work hour, increased (as did a host of problems). The limiting factor on field and farm size went away with the draft horse. Yet there are also some things we lost as a society when we discarded real horsepower, including self-reliance in the form of a farm-fed power source that could repair and reproduce itself; horse-handling skills; human-horse teamwork, communication, and relationships; and homegrown fertility in the form of horse manure. Feedstores, veterinarians, harness makers, blacksmiths, farriers, and other horse-based businesses that helped fuel a rural economy also went. We are not naïve enough to think that people would ever or should ever go back to a horse-powered agriculture, but we appreciated that we could have a glimpse into this world of humans and horses depending on one another.

Living and working with animals have helped us realize our own anthropocentric arrogance. Instead of thinking that animals are like humans, we recognize that humans *are* animals. As different species, we all have different abilities in terms of thinking and feeling and communicating. We also admit that we don't know and can't possibly know what is going on in their minds. We barely understand our own minds, so it's the epitome of hubris to think we understand theirs. We do believe they feel love and fear, have dreams, and solve

FIGURE 2.7. After adjustments by our employee Peter, John discs a newly plowed field with Nora and Nellie.

problems, as we do. We are not so different after all. We believe that this ecological disconnect of separating ourselves from nature, from other beings, from other humans, is the root cause of many of our societal problems.

Seeds of Change

Within about a dozen years of starting our farm, we had set up a draft horse–powered, diversified small-scale organic livestock and produce operation. We had worked out systems to integrate and take advantage of the synergies livestock offered to our vegetable-production practices. While management intensive grazing was regenerating the pasture, the manure collected from the animals during the winter was composted and used to improve the fertility and organic-matter content of the vegetable fields. We also incorporated crop rotation and cover-cropping in the vegetable fields and pulled the chicken and rabbit tractors down the rows of cover crops or post-harvest fields. This provided daily fresh forage for the chickens and rabbits while they added their manure to the soil. We were relatively early adopters and promoters of grass-fed meats, and we were garnering attention from the media and customers, which led to increased sales. We paid our employees more than the minimum wage and were able to offer them free housing. We felt as though we were doing everything to the best of our abilities and with the best intentions, yet we weren't quite satisfied.

We'd also run the gamut of marketing approaches. We started a farm stand at the farm and a farmers market at the nearby Smugglers' Notch Resort in partnership with our farming friends David and Jane, from River Berry Farm. We all started farming at the same time and have been lifelong friends and supporters over the years, a priceless relationship that has been an important factor in our success. David and Jane gracefully bowed out of these operations after a couple years to focus on their own retail business, but they continued to sell us wholesale veggies and bedding plants to help supplement our farm stand and farmers market offerings.

We sold produce to local grocers and restaurants and our meats directly to customers through an order-your-own lamb, pork, chicken, or turkey CSA program. We even had a small, twenty-four-family member CSA weekly produce box share. We loved interacting with and educating the local community and watching their families grow up on our food and recipes.

To our customers, it may all have seemed perfect, but there were inherent problems, including the economics of our small organic farm. It felt as if we were conducting an orchestra with all the correct instruments, but the music wasn't quite right, no matter how hard we conducted, and we were conducting like crazy. John was working really hard on the farm, sixty to eighty hours per week, with help from Nancy and the kids, but we didn't have a lot of net farm income left after paying for our employees and the necessary supplies and equipment to run a farm. There also wasn't any margin for error, especially with Mother Nature's reminders about who was in charge.

In other words, we weren't economically resilient, and while we were paying our employees above minimum wage, we couldn't pay ourselves that. One summer, John figured he made about half of minimum wage per hour, but luckily, he worked eighty hours per week to make up for it! Though we were feeding our family our own meats and vegetables and getting tax benefits to offset Nancy's income while we capitalized our farm with infrastructure and equipment, so that did make up some of the wage disparity.

We thought maybe we were doing something wrong, but when we looked around at our farming colleagues, we found many other small-scale vegetable farmers had similar stories. They, too, were young, energetic, passionate, and doing backbreaking labor with little economic return. Other small-scale livestock farmers were little better off. The organic vegetable farms that were doing better economically were in the forty-to-sixty-acre range and were input and equipment intensive with a larger number of employees and regional markets. The livestock farms that were doing better had even more acreage and larger numbers of livestock. With our small land base and commitment to the "Small Is Beautiful" concept, we weren't enthralled by or inclined to adopt those midsize models.

We began to observe slight wear and tear on our soils, too. From soil tests, we could see that organic-matter percentage was better, thanks to our cover-cropping, rotations, and additions of compost and manure, but the soil structure was changing. In certain areas, we noticed changes in the tilth or fluffiness of the soil or evidence of cracking of the soil surface after hard rains. We attributed this to our main weed-management program of rototilling, cultivating, and hoeing. This constant manipulation of the soil was burning up some of our hard-earned organic matter, wreaking havoc on macroinvertebrates such as insects and worms, and reducing the amount of humus and organic acids that hold the soil together in aggregates. These were slight changes noticed by our inquisitive nature, but they made us wonder what shape the soil would be in twenty-five years later.

Our lives were changing in other ways during those middle years. Our kids were growing up, graduating, and moving on. John became more involved in consulting and teaching organic practices during the off-season in such places as Haiti, the Dominican Republic, Guinea, and India. Both of Nancy's parents died during those years, experiences that reminded her and the family of the realities of impermanence and our own mortality. Nancy also started writing and making art in her spare time. As a professor at UVM, she could take courses for free. Working on homework (especially from her art classes) was an activity she could do with the kids. They would all sit around the kitchen

FIGURE 2.8. Our weekly CSA pickup at the farm gave us a chance to connect with the local community.

table, working on homework or projects together. While we were still committed to the farm, other things were vying for our attention.

Nancy's off-farm university job was still paying for most of the household bills, health care, and the mortgage. Even though Nancy's retirement from the university was still in the future, we knew we needed to start planning for the time when the farm would be our sole financial support. We also knew that climate change issues would only be getting worse in the years to come. Our small farm needed to become more economically and ecologically resilient. The only question was how best to do it.

FIGURE 3.1. Blueberry bushes provide a beautiful backdrop in the fall. The cow's skull may or may not focus cosmic energy, but it does serve as a reminder of the importance of animals in our personal food and farming ecosystems.

CHAPTER 3

pathways to resilience

Maybe it was the recognition that the life-support systems on the planet were really showing the wear and tear of human mistreatment, or our own growing restlessness and discontent with modern agriculture—whatever it was, a lot of ideas began to bubble up and catch our attention in the early years of this new millennium. Terms like "resilience," "agroecology," and "permaculture" were being touted as approaches to deal with the world's twenty-first-century agricultural challenges, including soil loss and degradation, climate change, water and air pollution, loss of biodiversity, and nutritional deficiencies and contaminants in our food. We began to immerse ourselves in these topics through reading, researching, and teaching at the university. We'd always seen the farm as an ecological system connected to the surrounding community and ecosystem, but we wanted to be even more proactive in having a beneficial relationship to the land. We wanted our practices to support the movement toward regenerative farming livelihoods and social justice, creating a space where people could learn and renew themselves spiritually while helping to make the world a better place. A challenge to ourselves.

There were also a lot of energetic, young organic vegetable and livestock farmers sprouting up locally, which was exciting but also posed a healthy competition to our livestock and veggie operation. This prompted us to ask how we could achieve more economic resilience. We noticed small entrepreneurial beverage companies were booming, including wineries, breweries, and cideries, some of which were just down the road. Our love of and interest in fruit, from our graduate school years and work in the fruit-growing area of Western Michigan, were still strong. This fruit connection had started even earlier in Nancy's case. She grew up helping her parents run the family fruit market in Western New York. Fruit was in our past, and we were beginning to realize it was in our future too.

In the following sections, we outline many of the topics, concepts, and movements that have helped nudge us into our current direction by informing much of our thinking and practices. We see many similarities and overlap in these movements and philosophies in that they envision food-producing systems that are more environmentally sound, economically viable, and socially just than industrialized agriculture (including industrialized organic) as it is now practiced.

Resilience

Resilience is the ability of a system (including a person) to bounce back from stresses and shocks. In ecosystems, resilience is enhanced by high biological diversity. High biodiversity reflects high numbers of species as well as the abundance of individuals of each species. An attribute of a resilient ecosystem is redundancy, where different types of plants and animals often share similar ecological functions (or niches). No two species have the same exact niche, but there can be lots of overlap or redundancy. If one species struggles due to climate fluctuations or parasites or pests, for example, another species can fulfill the ecological functions of that stressed species. The different plants and animals themselves may also have a wide variety of strategies and adaptations or even genetic variability within a population to make them more resilient in the face of change. One example of stored resilience can be seen in the form of seed banks in the soil from previous years. These seed banks allow plants to come back quickly after a stress or shock that may have hurt the existing population. This is an adaptation every annual veggie farmer has to face when the annual "weeds" come back after cultivation or soil disturbance.

Over the years, we've been learning how to deal with stressors such as climate change, new pest dynamics, competition, and changes in the marketplace. Our own aging bodies have also become a factor to consider as this has affected our energy levels and physical productivity, although we also like to think that it has led to increasing our wisdom! These uncontrollable factors have all been important and informed our choices over the years. Being small enough to be opportunistic and flexible has been an important survival strategy for our farm and for us. Resilience is survival.

To be able to cultivate ecosystem resilience, we've spent more time learning about the landscape, the soils, the watershed, and other features of the land as well as climate, weather, and the human development changes that impact the land. It's meant diversifying crops and the farm ecosystem. For us as individuals, it's meant a transformation in our thinking that led to the development of our perennial polyculture farm.

While we have always understood the importance of community, cultivating social resilience has become an expanding topic of study for social and agroecological researchers worldwide and a new area of emphasis for us. Social resilience applies to individuals and groups such as families, businesses, communities, and even countries. In times of stability, having social resilience might mean thriving and being more productive. In times of disruption or disaster, it relates to the ability of the individuals within a group to adapt, reorganize, and survive or even grow in response to the instability. To us, cultivating social resilience includes a range of strategies, from strengthening community ties to learning from peers to developing new market networks for economic security.

Social issues are invariably tied to food and personal security, environmental and land use issues, and other economic themes. Problems such as climate change, increasing economic disparity between rich and poor, overpopulation, and overconsumption are causing ecological degradation and scarcity-induced wars. The capacity of communities to deal with these stressors and recover, as well as to prepare ways to avert future disasters, may well mean the difference between death and survival for millions of people.

Organic Farming

The term "organic farming" arose in the mid-twentieth century, defining an alternative farming approach to the increasing modernization of agriculture with its use of synthetic fertilizers, monocultures, and pesticides. The foundation of the early organic movement focused on the tenet that healthy living soil equals healthy food equals healthy people. For us and many small and medium-sized organic farmers today, this is still our core philosophy. During the 1970s and 1980s,

TABLE 3.1. Stressors impacting The Farm Between with corresponding resilience strategies.

Category	Stressor	Resilience Strategies
Environmental	Flooding	• Planting flood-tolerant perennials in frequently flooded fields • Restoring riparian zones and protecting grassy waterways to slow down and soak in water • Decreasing soil erosion with no-till
	Drought	• Increasing organic matter and mulching perennials • Collecting rainwater from roofs
	Winter temperature swings / late and early frosts	• Choosing cold-hardy perennials • Planting hedgerows as windbreaks • Planting fruit trees in hoophouses • Overwintering nursery plants in the barn
	Pests and diseases	• Intensifying plant biodiversity to increase predaceous insects and birds • Growing disease-resistant varieties • Increasing biodiversity through soil health • Keeping plants dry in hoophouses • Screening for pests
	Pollinators in peril	• Eliminating pesticides • Increasing nesting, overwintering, and floral resources • Reaching out and giving presentations to encourage others to do the same
Social	Finding and keeping employees	• Working with students in universities and colleges and new-farmer programs to find interested workers • Paying well and keeping it fun and interesting
	Competition	• Growing uncommon crops and a wide diversity of crops • Staying fresh and relevant by growing and developing new products • Focusing on regenerative growing practices for a marketing edge
	Changing markets	• Networking to develop connections with local entrepreneurs and businesses • Staying small and diverse to maintain flexibility • Keeping up with trends
	Networking with communities	• Keeping active and volunteering in local, statewide, and other organizations • Operating a retail nursery and hosting public activities on the farm • Posting social media news and events

TABLE 3.1 (*continued*)

Category	Stressor	Resilience Strategies
Economic	Paying workers livable wage	• Providing seasonal housing, a decent starting wage, yearly wage increases, and other perks
	Paying ourselves	• Being cognizant of enterprise financial analyses to pursue economically viable activities • Being efficient with our employees so there is profit left over for us
	Maintaining infrastructure	• Allocating time and money for deferred maintenance projects on house, barns, and equipment
	Transitioning the farm	• Understanding that we can't do this forever, nor do we want to • Maintaining economic viability and infrastructure • Paying attention to other farms' transition strategies • Talking to young farmers
Personal	Aging bodies	• Eating healthy, hydrating conscientiously, exercising (walking, practicing yoga), maintaining a good sleep schedule, taking naps (John)
	Burnout	• Vacationing in the summer (a new concept for us!) • Maintaining and investing in other interests • Taking care of ourselves (see "Aging bodies" above)
	Keeping up with technology	• Picking and choosing the most advantageous technologies for us (for example, we don't want cell phones, but we use a credit card reader on our tablet for nursery sales; we also advertise on social media) • Getting help from people more in tune with changing technologies and trends

many states began regulating and certifying organic farming practices. While this was beneficial within a given state, it still meant variable standards and practices across the country. In 2002, the USDA National Organic Program (NOP) came into effect and set uniform nationwide standards for farmers to follow and certifiers to verify.

We have always farmed organically, and have been certified by Vermont Organic Farmers (VOF), which works under the auspices of the Northeast Organic Farming Association (NOFA) of Vermont. In 2002, when the NOP came into effect, we surrendered our certification in protest and stopped using the term "organic" in our marketing for a few years, although we still farmed according to VOF standards.

We felt that the USDA had usurped the term, and that the law resulted in a loss of local control in setting appropriate standards for practices. It also created an easy pathway for the industrialization and commodification of organic agriculture.

While some people argue that the phenomenal growth of organic product consumption in recent years is due to that law making large-scale marketing easier, we think it has also promoted a deviation from the original significance of organic certification. The original movement was informed by small and medium-sized farms. Now it favors large farms and an industrial organic mind-set and corporate greenwashing. Lobbyists and moneyed interests have been able to influence production standards, and corporations have jumped into the marketplace with increasing shareholder profits as their main motivation, a far cry from the intent of the original organic movement.

Because of this, organic farming is now at a crossroads. One path leads toward healing, and the other path toward destructive practices. Large-scale industrial organic stays clear of genetically modified crops and synthetic fertilizers and pesticides, but the organic standards do allow for certain naturally based broad-spectrum pesticides with human and environmental health costs, soilless-growing hydroponic systems using nutrient solutions, large-scale concentrated livestock operations, and large-scale monocultures. These organic farming systems show the same characteristics as conventional agriculture. They're not sustainable—and forget about regenerative.

What started out as a philosophy and movement away from conventional agriculture has broadened to include space for mimicry of the hallmarks of conventional agriculture, such as input substitutions, large monoculture plantings, and poor working conditions. Many corporations with both conventional and organic products have moved into organic food production to increase profits and not necessarily because of a foundational belief in the benefits of organic agriculture. We don't relish bashing industrial organic, but its trajectory is not the regenerative agroecological revolution the world needs, and we believe it should be called out.

We did eventually become certified organic again. We found "organic" could be a magical, feel-good term to the general public, and after getting asked, "Are you organic?" a million times, we eventually softened (or caved). Recertification was also a marketing decision to better serve our wholesale customers, who wanted to be able to put CERTIFIED ORGANIC FRUIT on their labels. Being certified organic also made it easier to attract direct market sales at farmers markets. If you don't have the time or inclination to try to explain your production practices to every customer, we found that "Organic Snow Cones" has much better cachet than simply "Snow Cones."

The Rodale Institute now promotes an add-on label of REGENERATIVE ORGANIC CERTIFICATION to the regular CERTIFIED ORGANIC label. The requirements to attain this add-on label involve more stringent standards in the areas of soil health, animal welfare, and farmer and worker fairness. We would also like to see a "no spray" practice included in the regenerative standards. Many organic farmers use NOP-approved chemicals that also have detrimental effects on the environment; for example, copper fungicides and broad-spectrum "natural" insecticides are toxic to nontarget insects such as pollinators.

We should mention that there is a parallel movement to regenerative organic, which advocates creating a REAL ORGANIC add-on label, led by a group of disillusioned organic farmers who were initially motivated by the failed fight against the inclusion of hydroponics in the National Organic Program. The USDA also recently relaxed rules requiring organic laying hens to have actual access to the outdoors. This whittling away of organic rules by the USDA and special interests has a lot of people wondering what the future of "organic" is.

While we see the importance of some type of label to differentiate ourselves from industrial organic in the eyes of consumers, in an ideal world, people would take the time and effort to know their farmers and how their food was produced, making labels obsolete. For now, we prefer the term "regenerative organic," but when marketing our farm and products, we often use even more descriptive terms for our approaches—for example, "no spray," "pesticide-free," and "biodiversity-based."

Regenerative Agriculture

The Rodale Institute, an educational and research nonprofit based in Pennsylvania, first began using the term "regenerative agriculture" in the 1980s. It was also on the forefront of the organic movement back in the 1940s. "Regenerative agriculture" describes farming practices that can slow climate change by rebuilding soil organic matter. The regenerative agriculture label also promotes biodiversity, improving water quality and enhancing ecosystem services. What's not to love? We support this term because its meaning goes beyond the current application of the term "sustainability," which seems to focus more on maintaining the status quo without doing further harm. We need to do more than that! Both the Earth and human society need an agricultural movement that is restorative and promotes healing, thus the term "regenerative."

In 2014, the Rodale Institute published a paper titled "Regenerative Organic Agriculture and Climate Change," which promoted the idea that wide-scale adoption of practices such as reduced tillage, cover-cropping and crop rotation, and the use of compost for fertility could sequester more than 100 percent of the world's annual CO_2 emissions. Concern over climate change is bringing more attention to the fact that with improved management practices, soil can sequester CO_2 and hold more water, therefore mitigating their impacts as greenhouse gases.

Concerns over the loss of biodiversity, such as declines in insect biomass and collapsing wildlife populations, are bringing to light alternatives to pesticides and chemical fertilizers. Agroforestry, no-till planting, perennial polycultures, management intensive grazing, and other regenerative practices are beginning to gain more traction in certain agricultural circles. Issues related to worker fairness—including safe working conditions and fair pay, which need improving in all agricultural sectors across the globe—are also addressed by the regenerative approach.

Agroecology

The term "agroecology" first appeared in the early twentieth century as a scientific discipline, primarily as an application of ecology and ecological approaches in the study of agricultural systems and crop production. This scientific approach combines ecology, biology, microbiology, agronomy, plant physiology, and several other disciplines, making it an integrative study of agriculture. More recently, agroecology is considered as a science, practice, and movement. The term has moved beyond agricultural systems to also include overall food systems with an emphasis on the social component. The basic premises of the movement are that the current food-producing systems are not sustainable for rural communities, and we need to change to adapt to the times we live in.

On the local level, these agroecological movements have become important catalysts for initiating practices and regenerative farming methods that focus on improving water quality and soil health, increasing biodiversity and plant and animal conservation, and sequestering carbon, to name a few. These practices are also concerned with improving farmworker compensation, health, safety, and community food security. "Food sovereignty" is a relatively recent term from the agroecology movement, coined in 1996 by members of La Vía Campesina, a peasant movement organization. La Vía Campesina defines food sovereignty as "the right of peoples to healthy and culturally appropriate food produced through ecologically sound and sustainable methods, and their right to define their own food and agriculture systems." We can't claim to be peasants. Or maybe we can and proudly so, because the term originally meant "country people." In any case, we certainly support this sentiment.

In conventional and even many large organic agricultural systems, the ecosystem services for soil fertility and pest management come from purchased inputs, such as fertilizers and pesticides. Our agroecological goal is to substitute synthetic and off-farm inputs with biological diversification and intensification on the farm in order to attain the same production benefits while supporting the overall ecosystem. For example, this also benefits our native pollinator, insect, and other wildlife populations. In other words, unleashing the power of biodiversity through biological inputs!

Over the years, we have developed a fantastic relationship with the Agroecology Livelihoods Collaborative (ALC) at UVM. We host classes, student work groups, and international short courses based on agroecology. We are working on long-term projects to measure soil health and biodiversity. These interactions with students and other agroecologists to cocreate knowledge constantly energize and inform us.

Permaculture

Around the time we jumped on the agroecology bandwagon, we also started giving more attention to permaculture. John taught a permaculture class at UVM. Books like *Gaia's Garden: A Guide to Home-Scale Permaculture* by Toby Hemenway and *Permaculture: Principles and Pathways beyond Sustainability* by David Holmgren were informative and inspirational. Permaculture, or "permanent agriculture," focuses on integrative perennial-based systems that work with nature and the natural landscape to provide food and other products for local needs. This focus on systems, including agroecosystems and communities, overlaps with many of the agroecological practices discussed earlier.

We've embraced the core tenets of caring for the Earth and all its inhabitants, including humans, as well as the idea of taking only a fair share. Permaculture's twelve design principles also aligned with many of our ecological farming practices. For example, the principles dealing with observing and engaging in nature and its patterns, integration of food production with natural systems, and increasing diversity were ones that we were trained to do as ecologists, and continue to practice as farmers. Our farming practices align with the permaculture principles of developing systems that catch and store energy, obtaining a yield for one's effort, reducing waste, using renewable resources, and adjusting our farming systems so that they continue to function. There is always room for improvement with all of these, especially in terms of biodiversity and taking advantage of edges and the marginal areas on the farm. Both diversity and edges, or marginal lands, have ecological, social, and economic manifestations. The final permaculture principle has probably become the most important for us: creatively responding to change.

Agroforestry

Agroforestry is another concept that falls within the realms of agroecology, regenerative agriculture, and permaculture. Agroforestry relates to the intentional incorporation of trees and shrubs into a crop or animal farming system, or both, for the benefit of the environment, the community, and the farm. Examples of agroforestry include establishing hedgerows, growing trees for biomass or food crops, grazing animals in orchards (silvopasture), planting crops within orchard alleys (alley cropping), and enhancing and maintaining riparian forest zones next to streams and rivers.

While agroforestry techniques have been used throughout history and cultures for human benefit, recent emphasis has been placed on the ecological benefits (which also benefit humans in the long run) of preventing desertification and soil loss, reducing deforestation, improving water quality, and enhancing wildlife populations. In addition, recent emphasis on biodiverse food forests and drought-tolerant systems may help in food security issues as a result of climate change.

Biodynamics

The biodynamics movement came about through the work of Dr. Rudolf Steiner in the 1920s. The basic philosophy is that each farm or garden can be thought of as a unique and integrated organism. Biodynamic principles and practices encourage plants and livestock to be raised together synergistically and for farms to generate their own fertility through composting and cover crops. The emphasis on biodiversity and observation to attain farm-related goals relates well with our own principles. Biodynamic certification requires that the farm meet the requirements of organic certification as well as other biodynamic measures, some of which are based on preparations and practices that are often viewed with skepticism by many in the mainstream scientific community. We still have a lot to learn about biodynamics.

Rethinking Cultural Philosophies

During the early 2000s, we also became attracted to the Japanese aesthetic of *wabi sabi*, a philosophy that finds beauty and value in the impermanent, and the natural cycle of growth, death, and decay. Wabi sabi isn't about being sloppy or messy, but more about letting nature be nature. According to this aesthetic, nothing is perfect or complete; therefore, we can embrace the weathered and aging look in nature and in ourselves.

Wabi sabi also teaches us not to force nature into something she's not. While we realize that farming requires a fair amount of ecological manipulation, we wondered whether there were ways to reduce that. Did we need to keep so many places neatly mowed and trimmed? Wasn't it better to allow wildflowers (from dandelions to asters) to come back in the meadows and grow around the barns? Or leave the dead flower seed heads for the birds to eat in the fall? We've come to think of wabi sabi as not only an important aesthetic or philosophy, but an ecological view as well.

We were also becoming more interested in spiritual traditions that focused on respecting and caring for the Earth. In particular, we explored our own European heritages from ancient Ireland (John) and Scandinavia (Nancy) and connected with ancient practices and philosophies that focused on nature and natural cycles. Again, we found these resonated with our ecological view that human beings are part of nature, are nature in fact, and are not separate from it. How could we farm more in line with our philosophies?

It was during this time that we started thinking more about perennial fruits. We recently expanded our blueberry plantings and were enjoying the cherries and plums from our fruit trees we'd planted when we first moved on the farm. Why not do more? When we heard that the son of our dairy farmer neighbor down the road was starting Boyden Valley Winery, it planted the seed of an idea for growing fruit for this wholesale market. We were intrigued with black currants, a fruit that has both strong Irish and Swedish ties. We were awarded a Northeast Sustainable Agriculture Research and Education (SARE) grant in 2002 to conduct a black currant variety trial. A couple of years after that, we harvested the first black currant crop with our kids and their friends. We sold the fruit to our neighbor's winery, and the delicious Cassis wine they make is a favorite to this day. This larger-scale currant planting was really the beginning of the transformation of our farm's focus to perennial fruit, although we didn't quite realize it at the time.

During this period, we also started new enterprises and practices. We added several thousand-square-foot passive solar hoophouses. Initially, we used them for tomatoes, but within five years, we converted

FIGURE 3.2. Our kids (two homegrown boys and two adopted girls) help to plant our first black currants into a rye cover crop in 2002.

parts of them to fall-bearing raspberries. We also experimented with no-till production for our pumpkins by roller crimping a rye cover crop to kill it and transplanting the pumpkins into the thick residue, which acted as a mulch. The rye also kept the growing pumpkins clean until harvest. At that time, we also began implementing new ideas for pollinator plantings such as flowering groundcovers and cover crops that created floral resources. Agroecological, regenerative, and permaculture ideas had taken a greater hold on our minds and our practices, and it felt good.

FIGURE 3.3. Cherry tomatoes grew well in the hoophouses. They were a favorite at our CSA and local grocery stores.

Rewilding

Rewilding the farm has been an ongoing project for us. When we bought the place, we knew it was lacking in life even though it looked like the iconic, bucolic Vermont dairy farm. Every square inch was mowed or hayed except for a small wooded area in the corner of the back pasture. In those early years, we didn't have a tractor or a brush hog–type mower, so we replaced the mowing machines with grazing animals and stopped mowing many areas, especially patches around barns and the ditches. Reduced mowing was an easy first transitioning step. It made sense to us too. As kids, we'd love these types of wild places for exploring and hiding out in "forts." We also knew they were needed for grassland birds, such as bobolinks and meadowlarks, and for other wildlife. Why not make less work for us and create wild places for us, our kids, and wildlife?

After we stopped mowing and cleaning out ditches, shrubs and trees came back—and so did the birds. The birds didn't come back just because of the berries and shelter provided by the trees and shrubs, but for the insects too. Insects and insect larvae eat the leaves of native trees and shrubs. Birds, even seed-eating birds, need to feed insects or other arthropods to their young as an important protein source. It's all connected. We loved watching the barn swallows return in the spring and the song sparrows swoop into the barn when the door was open to

FIGURE 3.4. Baby robins share a space on the farm.

grab a few big barn spiders relaxing in their webs. They were going to make a tasty meal for their babies.

The increase in bird numbers and the overall insect diversity on the farm are what we have noticed changing the most over the years. We're so glad they've moved in. A few summers back, an avid bird-watcher (or maybe we should say "bird-listener") attended a farm tour and identified more than twenty-five different bird species in about half an hour from their songs and calls. Even in winter, the crows caw while chickadees, blue jays, and juncos chatter along the stream and the back-fence line with its hawthorns, ash, and other trees and shrubs. At the end of February and beginning of March, the male red-winged blackbirds come back to the farm, staking out their territory before the females return later in the spring. They're so cheerful in the morning with their *gurgle-lees* call and bright red wing bands. The yellowthroats and chestnut-sided warblers follow in May, and flitter in the ecotone between woods and pasture. Even our dogs are excited by the calls and songs of spring.

Letting our stream edges grow into birch, willow, box elders, and a variety of other trees and shrubs created a three-dimensional habitat while providing food sources for birds and pollinators. It also formed a nice visual screen from our neighbors. The people we bought the farm from had moved up to a nearby hill overlooking the farm. How could they help but watch what we did with it? The old owners didn't appreciate our new "scruffy horticulture" look, though. Through the community grapevine we heard various complaints about how we weren't keeping the place up. The truth was, we were "keeping it up" ecologically much better than in the past.

We'd been rethinking our predominant cultural aesthetic of manicured lawns and manicured farms since day one, but we still had a lot of mowed lawn. We asked ourselves why and couldn't come up with a good answer. We could only come up with more reasons to reduce the lawn to an area large enough to play badminton and have a lacrosse catch. That's when we decided to turn the front lawn into a small-scale apple orchard.

John had loved apple orchards and fruit growing ever since his graduate school days studying biological control of apple pests. Almost no one in Vermont was making the move toward growing organic apples, and only a few were growing organic fruit. The IPM strategies that he had worked on during graduate school twenty-five years earlier had not lived up to their promise of replacing pesticides as the mainstay of a pest management program. In fact, conventional and organic apple growers were often backsliding into the "regular calendar spray schedule" mentality whether there were pests present or not. Also, the pesticides had changed. Neonicotinoids, a new class of broad-spectrum insecticides had entered the market in the mid-1990s and were gaining ground, though evidence was growing that they were negatively impacting bee populations.

Organic fungicides, copper and sulfur, have negative environmental impacts as well. Additionally, broad-spectrum organic insecticides such as Entrust (manufactured by DowDuPont) are toxic to pollinators and other beneficial insects. We weren't interested in those models.

Organic apple production in the Northeast United States isn't widespread. Trying to grow blemish-free apples in a conventional orchard setup with organic sprays as substitutes doesn't work economically or ecologically. We wanted to grow no-spray organic apples and prove it could be done. We sorted apples for fresh eating and pressed gnarly ones for cider. We reinvented our side-front lawn by turning it into an orchard flanked with rows of unusual fruits (gooseberries, currants, honeyberries, and lingonberries). When we planted hundreds of trees and shrubs in other areas on the farm and let our

FIGURE 3.5. Late-summer asters (purple and white), goldenrod, joe-pye weed, and jewelweed came in when we stopped mowing edges.

FIGURE 3.6. Fruit for all replaced our large front and side lawn. Aronia is in bloom behind.

pasture revert to a wildflower meadow, we heard even more grumbling from our neighbors. Bucking convention can result in raised eyebrows by those who believe anything less than carefully manicured green lawn—that is, alien grasses that are near biological deserts dominated by a few species—and mowed fields looks unkempt.

But attitudes are changing. As each one of us begins to realize that lawns *are* part of the problem, we may see that reinventing our lawns, turning them into diverse food-growing and native plant gardens, can be part of the solution. This was done with victory gardens in both world wars and more recently with the Food Not Lawns movement. We like to imagine the change in nutrition that the typical American diet would undergo if lawns became gardens and food forests. We think about the increase in birds and other wildlife if backyards were filled with native biodiverse gardens. We think about biodiverse, organic, rich soils, and trees and shrubs sequestering carbon in millions of newly planted acres.

CHAPTER 4

it's all about the soil

Healthy soil is, and has always been, the foundation of our organic regenerative farm. Healthy and biodiverse soil regenerates itself through the interactions among the microbes, other soil inhabitants, and the root zone to create healthy plants and healthy food, while also storing carbon and cleaning water. That is one reason why allowing the organic labeling of hydroponic tomatoes or other hydroponic crops makes no sense to us. That is why using fungicides, herbicides, or pesticides (even organically approved ones) that get into the soil and kill untargeted organisms makes no sense to us. These chemical tools are best thought of as ecological sledgehammers that have detrimental effects on nontarget insects, soil fungi, and other beneficial organisms.

At our farm, we want to be purveyors of life, not death, and promoters of biodiversity, not the sterility of the monoculture mind-set. The founding principle for organic has been to "feed the soil." While special interests and Big Food may have usurped the term "organic" and are eroding its fundamental tenets, we will continue to march to the beat of ecological, regenerative, and biodiverse agriculture with special consideration for taking care of the living soil that we are a part of, and that is a part of us.

Soil Is Alive

Whether we've focused on raising vegetables, animals, or, as we do now, fruit, it's always been about the soil. And the scary thing is that since the 1950s, the Global Assessment of Land Degradation and Improvement estimates we have lost almost two billion hectares of arable land (about 22 percent) worldwide due to soil degradation. Our society treats soil like dirt, and as our friend Vic Izzo, an instructor at UVM, likes to say, "Dirt is dead. Soil is alive." We need to adopt a "living soil" cultural mind-set, one that acknowledges soil, the skim layer of biofilm resting on top of the parent material, as the support for all terrestrial life on the planet. Soil is the key to resilience and human survival!

In fact, humans and other organisms are all living manifestations of soil. We are the recycled carbon, hydrogen, oxygen, nitrogen, phosphorous, potassium, and other stardust elements that came together to form this planet Earth over four billion years ago. We are connected and a part of everything and everyone else that has existed through

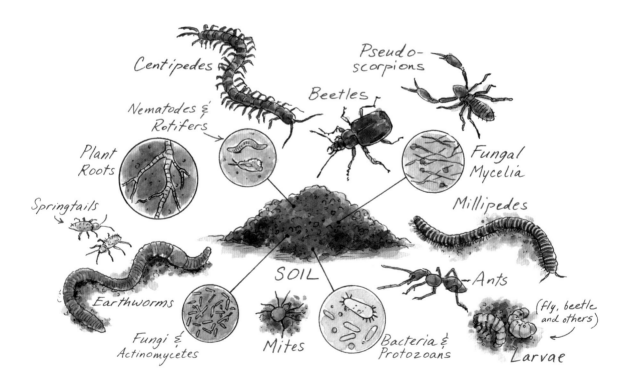

FIGURE 4.1. A few examples of the inhabitants that live in soil. *Illustration by Elara Tanguy.*

this common origin and pathway of the soil. Science has shown that certain microbes release calming endorphins when we plunge our hands into their soil home. Other microbes help populate that little pouch of soil-like activity that we carry around in our gut. We rejoice in being part of this dance of life with microbes, fungi, plants, humans, and other animals continually springing forth from the biofilm during a shared time and space. To be disconnected from the soil is unthinkable. We shall return to the soil and eventually become parts of new organisms as they take their turn in the dance.

Over the years, we've enjoyed learning about the scientific studies that examine many of the processes occurring among microbes, fungi, and plants, but we also recognize that the soil harbors infinitely complex interrelationships among all things. These interactions are beyond our capacity to completely unveil through scientific methods or models. It reminds us of our efforts to try to understand the infinite universe, in that we get glimpses, yet never full comprehension. Not that we should ever stop studying soil, but perhaps at times, we should recognize our intellectual limitations and sit back and revere soil as being part of the sacred mystery of life.

As ecologists and longtime organic farmers, when we reflect on soil and soil cycles of birth and decay, we think about all the organisms—the plant roots, bacteria, fungi, insects, and more—that call the soil their home. These are our partners and cocreators. It's estimated that a teaspoon of healthy, rich soil contains millions of organisms and that 95 percent of these soil organisms are unclassified and unknown. Much of what we know about ecosystems comes from aboveground and aquatic studies. Soil ecology remains pretty much undiscovered territory. While new DNA sequencing and profiling tools help provide estimates of genetic diversity, these tools don't help us discern the holistic community structures and relationships that are critical for understanding an ecosystem. After all, ecology is the study of communities and interactions.

Scientific studies are beginning to shed more light on the importance of fungi and their role in creating healthy soil and healthy plants. While we knew that fungi are the great decomposers of organic matter, recycling it into slow-release nutrients and carbon, we are now even more appreciative of the benefits of fungi-dominated soil systems. In mutually beneficial relationships, plant roots exude sugars as a reward for the fungi that make nutrients available from the soil. Fungal mycelia can extend over large spatial scales and act as communication networks within the soil. With their antibiotic properties, beneficial fungi can also help keep diseases and pathogens in check. No-till agriculture favors a fungal-dominated food web near the soil surface, which leads to increased aggregation of soil particles, increased decomposition rates, and overall greater soil carbon storage.

Besides living organisms, soil also includes organic matter, nutrients, and minerals. It includes the geological parent rock material from which it's derived, as well as the water and the air within pore spaces. We want our soils to act as a carbon and water sponge, not fully saturated but more like a wrung-out sponge. We can continually improve soils by adding organic matter in the form of compost, manure, and woodchips and by not working them over with tillage. This improves the tilth (structure) of the soil by increasing pore spaces between the particles, while the carbon from these inputs feeds the living system to allow for long-term fertility.

It's now widely believed that even if we could dramatically reduce greenhouse gas emissions, it would still not be enough to reverse catastrophic climate change. We must also remove carbon from the atmosphere. Sequestering carbon in soils by increasing organic matter and using no-till farming is an important way to do this. How we treat the soil has always been a matter of survival for civilizations. Ours is no different.

Physical, Chemical, and Biological Aspects of Soil Health

Increasing fertility, organic matter, tilth, and other beneficial characteristics of the soils on the farm has been paramount since day one. Healthy soil is the foundation of regenerative agriculture and our focus. Digging around in the soil, touching it, and noticing textures, color, aggregation of particles, moisture, and living organisms are great ways to learn about soils. Observation of the vitality of the growing plants can also provide a lot of information and has been one of the most important indicators of soil health for us over the years.

In early years when we were getting to know our soils, we occasionally sent samples to the UVM soils testing lab for chemical and percent organic matter analyses. In more recent years, agroecology students working on projects on the farm have taken samples. The student projects used a relatively new comprehensive soil health assessment test, offered by Cornell University. The test measures key physical, biological, and chemical aspects of soil samples. As an

TABLE 4.1. Cornell's soil health test results from The Farm Between's Knoll Orchard

Grower: John and Nancy Hayden

Sample ID: s719
Field ID: E
Date Sampled: 10/25/2018
Crops Grown: APP/APP/APP
Tillage: no till
Coordinates: Latitude: 44.647505000000
Longitude: -72.854610000000

Measured Soil Textural Class: **Medium**
Sand: —% Silt: —% Clay: —%

Group	Indicator	Value	Rating	Constraints
physical	Surface Hardness	134 psi	63	
physical	Subsurface Hardness	286 psi	54	
physical	Aggregate Stability	66.6%	97	
biological	Organic Matter	6.1%	99	
biological	Soil Respiration	1.2 mg	97	
chemical	Soil pH	6.4	100	
chemical	Extractable Phosphorus	10.8 ppm	100	
chemical	Extractable Potassium	147.7 ppm	100	
chemical	Minor Elements	Mg: 157.2 ppm Fe: 9.5 ppm Mn: 3.8 ppm Zn: 3.0 ppm	100	

Overall Quality Score: **90** / Optimal

Source: From the Cornell Soil Health Laboratory, Department of Soil and Crop Sciences, School of Integrative Plant Science, Cornell University, Ithaca, NY 14853. http://soilhealth.cals.cornell.edu.

example, the results for a composite soil sample the students took from our Knoll Orchard are re-created in table 4.1, with the various soil characteristic ratings on a scale of 0–100.

The physical characteristics of surface hardness (0–6″ / 0–15 cm depth) and subsurface hardness (6–18″ / 15–46 cm depth) are measurements of soil compaction. This relates to the size of pore spaces and the ease with which roots, fungal hyphae, and other organisms can move through the soil. Compaction is measured in the field with a penetrometer, and expressed in the number of pounds of pressure per square inch (psi) required to push it through the soil. Our untilled, silty soil in this location is a little tough to push through, especially during the drought we were experiencing, hence the lower ratings.

The other physical characteristic reported is aggregate stability. This measures how well the soil holds together under a simulated rainfall event. Aggregates are also an indicator of biological activity as soil particles are held together by exudates from microbes and other soil inhabitants. Sticky soil is good, hence the excellent rating for this location.

Soil pH is a chemical measure of acidity on a scale of 1–14, with the lower end being acidic and the higher end basic. Seven is neutral. When we tested the pH of our pasture soil twenty-five years ago, the pH was around 5.5, on the acidic side for pasture. This prompted us to add dolomitic limestone to the fields. This agricultural limestone contains calcium and magnesium, important minerals for healthy soils and plant growth. Agricultural limestone also buffers the soil by reacting with excess acidity (H^+) in the soil, thus raising the pH. The Cornell results show that our soils in this location currently have a pH of about 6.4, which is ideal.

The other chemical measurements from the Cornell results are for phosphorus, potassium, and minor elements needed for plant growth (all measured as parts per million, ppm). These measurements all have very good ratings at the sampling location in table 4.1. In fields for vegetable growing and now for most of our fruit, we've regularly added compost or aged manure. This increases organic matter as well as important nutrient elements such as phosphorus. Our compost and aged manure tend to be acidic in nature, so we also regularly add lime or wood ash. Not only does wood ash neutralize acid, but it is also high in potassium. The only fruit whose soil does not receive wood ash or lime are blueberries and lingonberries, as both require a low pH soil (4.5–5.5) to thrive.

The biological soil health indicators used by Cornell include percent organic matter and soil respiration; the latter represents the amount of carbon dioxide given off by the sample over a set period of time. Both measurements for this soil had excellent ratings. Fred Magdoff, professor emeritus in the Department of Plant and Soil Science at UVM, often referred to organic matter as "the living, the dead, and the very dead." All organic matter is in different stages of decomposition. Increasing the amount of organic matter in soils is critical for several reasons. It increases the water-holding capacity of the soil, which helps plants when rainfall is sparse. Organic matter is also a critical component of the food web of the soil, slowly releasing nutrients as the soil organisms consume and excrete it in various forms, reproduce and die, and decompose ad infinitum. The organic carbon cycle is intimately tied with the release of nitrogen, often in the form of ammonia and phosphorus, which is bound up in the organic materials. The application of synthetic nitrogen fertilizers in conventional agriculture has decoupled the carbon and nitrogen cycles. Nitrogen can now be added without the addition of carbon, which results in lower organic-matter content and poor-quality soils.

Organic matter also enhances soil structure by allowing for good aeration in the root zone. Through their rooting, plants take advantage and expand this biologically active zone. A complex ecosystem thrives in soils rich in organic matter because of the diversity of niches the organic matter provides, not only in its physical structure but in its biochemical diversity as well.

Soil organic matter, like the ecosystems it supports, defies complete human understanding or modeling. It is a complex mixture of many kinds of long-chained carbon compounds with attached functional groups containing oxygen, nitrogen, sulfur, and other essential elements. We attempt to classify organic matter into such categories as humic and fulvic acids and humus, based on the length of carbon chains, solubility, and other properties, but organic matter is much more than its scientific classifications. The different chemical and physical structures provide the diversity of micro-niches that support the millions of different species living in the soil.

Increasing organic matter in the soil, whether in the hoophouses, the vegetable garden, or perennial fruit plantings, has always been an important strategy for increasing plant health and soil water-holding capacity (and therefore drought tolerance). The chemical structure of soil organic matter makes it hydrophilic (water loving) and able to chemically retain water. Its physical structure creates the micropores that trap water through capillary action. Increasing water retention by increasing organic matter content is only one of the benefits of increasing soil organic matter. Recent soil analyses done on the farm show that over the years, we've been able to more than double the organic matter in our soil from 2.5 percent to over 6 percent in many spots.

Another important component of the soil is clay, which affects water-holding capacity, water movement, and aggregation, as well as the exchange of positively charged ions in the soil. As with organic matter, these factors provide chemical and physical diversity in the soil, which affects biological activity and diversity. Heavy soils, which tend to retain more water, often have significant percentages of clays that affect what plants can grow there. We feel blessed (and at times, daunted) to have a diversity of heavy, medium, and light soils on these eighteen acres of land that we call home.

Soil Biodiversity

Photosynthesis and decomposition are two major life-generating processes. Aboveground, we are able to witness photosynthesis, plants' ability to turn carbon dioxide into carbohydrates and more plant material by using the energy of the sun in their green leaves. The plants release oxygen as a by-product, which allows the rest of us aerobes to breathe and live. Just as important but what we don't often notice, is decomposition. In the soil, where sunlight doesn't penetrate, the soil organisms busily consuming plant residues are the decomposers. Many also use oxygen in their metabolism and release carbon dioxide in the process.

While plant roots and some soil insects can be seen with the naked eye, most arthropods, nematodes, fungi, and microorganisms require the use of microscopes (often high-powered microscopes) to locate their presence. Biological activity in soil is often determined by measuring the amount of carbon dioxide respired by the soil organisms. That's right, the soil is breathing! The results in table 4.1 and other recent samples taken on the farm indicate high soil respiration in our farm soils. That's exciting.

The plants themselves create a diversity of ecological niches because of the physical, chemical, and biological nature of their root systems. For example, taproots and fibrous roots create physical variations. Plants impact the soil chemistry through their root exudates and influence moisture, oxygen, and nutrient contents in the rhizosphere (root zone). Different types of plants specifically influence the presence of nitrogen-fixing bacteria and other microorganisms. Above ground, plants interact with other plants, provide shade, and provide food for herbivores.

Soil organisms are the recyclers of plant materials, converting the stalks, leaves, and roots into that complex biochemical material called soil organic matter. They use lots of oxygen in this process and release carbon dioxide. There are a variety of microorganisms that utilize ammonia (NH_3) as their food (energy source). They are called nitrifiers and convert ammonia into oxidized nitrogen, primarily nitrate. Others use nitrate when oxygen is low (denitrifiers) and produce nitrogen gas.

There are also microorganisms that take nitrogen gas from the air (nitrogen-fixing bacteria) and convert it to compounds (nitrate, for example) that can then be used by plants. These organisms live in a symbiotic relationship with their plant host. Annual legumes including vetches, peas, and beans and perennial trees and shrubs such as black locust, honey locust, and Siberian pea shrub are good examples of nitrogen-fixing plants that we encourage on the farm.

Like us, many of the soil bacteria and fungi and other soil creatures need oxygen for their respiration. They release carbon dioxide into the air as a by-product of respiration. During tillage, soil is aerated. This increases the soil organism metabolism of the organic matter. The microbe population explodes and uses large amounts of soil carbon. This oxidation of soil organic matter happens every year when farmers across the country and across the globe plow their fields (thereby aerating the soil) for growing annual crops such as corn, wheat, soybeans, or vegetables. The loss of soil organic matter through microbial respiration adds a tremendous amount of carbon dioxide into the atmosphere each year, as well as depletes the soil's essential organic matter.

The loss of carbon and potential soil loss from erosion due to the soil's surface being exposed to wind and rain during annual grain production is why no-till and perennial systems are gaining interest worldwide. No-till reduces the oxidation of organic matter, stores carbon, and creates a fungal-dominated system that is beneficial to the plants. No-till systems sequester carbon and can create long-term sinks for carbon dioxide, unlike annual cornfields. Unfortunately, the predominant way to kill cover crops and weeds in no-till systems is by the use of glyphosate-based herbicides (such as Roundup), which is under scrutiny as a potential carcinogen and for its role in many other health problems. Recently it has even been shown to negatively affect the gut biome in honey bees. Let's not forget that we are all connected.

Composting

If we had to rank our strategies for building soil health, composting would be number one. Rarely a day goes by that we don't make compost by collecting food scraps, add to one of our piles, use compost in our fields and gardens, or talk about compost. Why? Because compost is critical for building up organic matter and feeding the soil. Making compost is both a science and an art. A science because we know a lot of the necessary requirements, biochemistry, and biology involved. An art because there's so much we don't know, and things don't always go as planned.

We relish taking in organic materials from the community (for example, horse manure, leaves, cardboard, shredded tax returns stockpiled by our accountant, and at times food scraps) and turning them into this essential part of our farming system.

Composting is the biological process whereby plant residue, manure, and other organic materials are transformed to more stable (less biodegradable) organic matter (humus, and humic and fulvic acids) by decomposers. During this process, the organisms use oxygen for their metabolism and release carbon dioxide, just as we do to digest our food.

On the farm we employ several types of composting to achieve different results. Our go-to method for larger-scale production is thermophilic, or high-temperature, composting. This method encourages heat-loving microorganisms that thrive at high temperatures. We use this method when we want to kill potential pathogens and weed seeds and make a lot of compost fast. We can encourage the microbial activity (zillions of little creatures doing their thing) by providing a carbon-to-nitrogen ratio of about 30:1 or 40:1. The mixing of stock materials to achieve this ratio has become intuitive over the years, and like a chef we will add a pinch (or wheelbarrow load) of more carbon, such as woodchips, straw, and shredded paper, or more nitrogen in the form of manure, apple pomace after pressing, green weeds, or food scraps to get a proper mix as we build the pile. Unlike chefs, we don't taste it.

In the early days on our farm, when we had a lot more energy and no tractor, we turned our thermophilic compost piles with a pitchfork, often with the help of our seasonal workers or lacrosse players on Coach Appreciation Day. We then loaded compost or aged horse manure (also by hand) into our ground-driven horse-drawn manure spreader and applied it to our annual crop fields with our team of mares. The one drawback of being a horse-powered farm was that horses don't have a front loader bucket. In conjunction with our horses going into retirement and us getting well into our fifties with the associated aches and pains, we did away with the bottleneck of pitchforking compost and bought a 27-horsepower tractor with a bucket loader. This allowed for an increase in our composting capacity, which we needed for our new fruit plantings and nursery business.

FIGURE 4.2. John's lacrosse team members are "cross-training" by creating a compost pile from winter pig bedding during the middle years.

FIGURE 4.3. Turning compost with a tractor has allowed for larger-scale production during the later years.

Nowadays, once our compost thermometer tells us that the pile is starting to heat over 130°F (54°C), we turn it with our tractor bucket to aerate and to mix the edges of the pile into the center. We turn it three or four times over the period of a few weeks to keep it thermophilic, then let it rest for another four to six weeks. At first, the temperature climbs too high for arthropods, worms, and many organisms to survive or thrive, but after the thermophilic phase, the compost cools down and arthropods and other organisms recolonize the pile. These organisms help increase the conversion of organic material and nitrogen compounds to more stable end products.

To encourage a fungal-dominated compost, we also employ a slower, cooler composting method without turning and adding oxygen. This recipe uses a higher carbon-to-nitrogen ratio (at least 100:1). Woodchips or sawdust are the main ingredients, with a little bit of nitrogen-rich weeds or manure added. Organic material conversion still occurs, but at a slower rate compared with the high-temperature methods. At the relatively cooler temperature, thermophilic bacteria never get started. As the material breaks down, we notice the white strands of fungal mycelia spreading through the pile in search of more carbon. This kind of composting can take over a year and it may not be as effective in killing off all the pathogens and weed seeds, but our trees and bushes seem to love it. It takes less effort but more patience, and we are trying to cultivate that.

Vermicomposting, another composting technique, takes advantage of worms that graze on fungal and organic materials to convert the raw organic materials into compost that is rich in humus and nutrients. We call the resulting product "black gold." Vermicompost systems are loaded with lots of other little critters like mites, springtails, and sow bugs, but dominated by the red wriggler worms that give it its name. These surface-feeding worms graze on the fungus and other bacteria growing on the rotting food as well as dining on a bit of the rotting material itself.

Our basement worm-composting operation consists of several ten-cubic-foot composting bins that we load up with leaves, shredded paper, or hay for bedding and our own vegetable scraps for worm food. We continually add thin layers of vegetable scraps over time. After the worms have worked their way halfway up the bin, we start emptying from the bottom. The worm compost is usually applied to our nursery plants at the beginning of the season. We will also make a worm compost tea for watering our potted plants in the nursery.

In each of the different kinds of compost we make, there exists a complex food web of predators and prey, a fascinating spectacle for microorganism enthusiasts. When the university gave away its decades-old scopes, John managed to be there at the right time and get one. A stereomicroscope allows for three-dimensional viewing. It's not as powerful as a light microscope but perfect for insect work and looking at soils. We love looking at soil and compost through the stereomicroscope.

We like to imagine the hoopla and shouting of Antonie van Leeuwenhoek (the inventor of some of the earliest microscopes) when he first witnessed all those wriggling animalcules in a drop of pond water. Or maybe he didn't say anything, his jaw dropping and mouth wide open as he looked. His heart must have pounded in an adrenaline rush as he moved his view around the drop, grabbing more slimy rotting vegetables and soil to view. He just couldn't stop. We know that feeling of excitement when we see the variety of crawling and wriggling "animalia" in our compost. They each have a story.

Some of our favorite compost and soil creatures are pseudoscorpions, also known as false scorpions. Their bodies are about the size of a sesame seed. As they roam the coffee-brown chunks of compost, they hold out their pincer-tipped pedipalps like knights holding out their lances. Pseudoscorpions belong to the same taxonomic order as spiders and scorpions and have eight legs. Like spiders and scorpions, they are also predators. In their case, they prey on springtails, proturans, and other little insectlike creatures called hexapods that also live in the compost. We see lots of them in our vermicompost. They're visible without the aid of a microscope, but with the stereomicroscope, they seem all the more menacing as they look for their next meal or maybe their next ride; hanging on with one of their claws, they are known to hitch rides on the legs of beetles or flies to aid in their dispersal. This ride hitching is called phoresy (biologists have a name for everything!) and it is pretty common.

With over three thousand different pseudoscorpion species worldwide, you'd think they'd be well known, but

FIGURE 4.4. Worms taken from the black vermicompost bins.

most people are oblivious to their existence. The same goes for most of the other soil-inhabiting creatures. Pseudoscorpions are secretive and small, which could account for their being overlooked. Or if someone does happen to see them in the bathroom or basement, they might mistake its identity for a tick and squash the little critter with their shoe. The irony is that pseudoscorpions probably eat ticks. That's because they are one of the top predators in the places they inhabit. They deserve the respect we give to other top predators such as lions or wolves. Those extra-large pedipalps that they hold out in front of their bodies are just as impressive as fangs. It turns out that because they live mostly in the dark, they've developed sensory bristles on their pedipalps that help them navigate their world. Their pincers also contain venom that they can inject into their prey. And if you're going to grab on to a beetle or fly for a free ride, you want a pretty hefty pincer and pedipalp so you don't fall off.

Close the Loop Chicken Coop

One of our favorite composting and livestock projects involved our laying hens. We called it "Close the Loop Chicken Coop," and it involved feeding food scraps to our hens and then thermophilically composting their leftovers and manure. With the help of a local solid waste initiative grant, we approached local schools and businesses and offered to take their food scraps if they paid a fee of ten dollars per thirty-gallon bin. Several signed on because they wanted to keep the scraps out of the landfill and liked connecting with local farmers to "close the loop." While the quality of the slop from grocery stores was variable (sometimes consisting mostly of coffee grounds or containing whole watermelons, which weren't that easy to handle or nutritious for the hens), the elementary school leftovers were great: a premium mix of PBJs, fruit, yogurt, turkey casserole leftovers, vegetable soup, you name it. Chicken heaven! We know this sounds gross, but chicken nuggets were our hens' favorite food.

We were paid to haul these food scraps, and since there were plenty and a wide variety of food, we didn't need to feed the hens any grain; so essentially, the eggs we got from them were free. Composting the community's food scraps also kept them out of the landfills, where they would eventually have produced methane as they broke down. In the winter, when the hens were in the barn and not on pasture, we composted leftover food scraps mixed with bedding and chicken manure or fed it to our composting worms for more finishing. Our hens got fat and sassy. What could possibly go wrong?

it's all about the soil

Pressure washing all the gross containers was not fun, especially on cold winter days, but it wasn't that bad. At first, getting frozen food waste out of the totes was a pain until John figured out a way to lightly tap them with a sledgehammer all around, resulting in a big *thunk*. The hens didn't seem to mind the frozen food, which looked like tote-tower sculptures. Things were going great until that spring, when John noticed a few rats in the barn. They say if you see one rat, there are at least fifty you don't see. Well, then we started seeing more, a lot more. They were enjoying the free buffet, roaming the horse stalls, scurrying (and later, not scurrying) away when we came near.

The rats were getting bad, and when John tweaked his back lifting a 150-pound food-waste container while simultaneously spilling rotten soup all over himself, he came home, changed his pants, lay on the couch, and told Nancy he was done! A few days later, he contacted a young couple who were also raising laying hens in town and gave the business and accoutrements to them. Food-scrap collection is definitely a young person's business—a young person with a strong back and rat terriers—and it's been taken up by farmers in our area who own large flocks. Unfortunately, they have drawn the attention of state regulators and are now being required to register as solid waste disposal operators. We're glad we got out when we did.

Cover-Cropping for Soil Health

Cover crops are grown for the benefit of the soil rather than for their own harvest. Cover-cropping was a key strategy for soil health in our annual vegetable production. It serves many beneficial purposes, including weed suppression and providing habitat for pollinators and other beneficial insects, but cover crops' real superpowers are preventing soil erosion and improving soil health. Bare soil exposed to wind and rain invites erosion. Watching soil blow through the air or running off into a stream is discouraging to say the least. As it is estimated to take about five hundred years (or more) to make one inch (2.5 cm) of soil, and because many conventional farming practices can result in the loss of that much to wind and water erosion in a matter of years or even months, a green blanket of cover crops makes a lot of sense.

Cover crops also add organic matter to the soil by harvesting atmospheric carbon dioxide and using solar power to turn it into carbonaceous plant material. Cover crop plants such as buckwheat, rye, or clover eventually find their way back into the soil by dying and decomposing or by being plowed under. Cover crops can break up compacted soils through the action of their roots. Using deep taprooted forage radishes, for example, is a great way to break up a compacted hardpan. When the roots decompose, holes for water penetration remain. Cover crop roots can also scavenge nutrients by drawing them back from where they have leached deeper into the soil back up to the surface as growing plant material. Certain legume cover crops, such as hairy vetch, alfalfa, and clovers, can team up with rhizobial bacteria to fix atmospheric nitrogen and make it available to plants in nitrate forms, sometimes at over one hundred pounds per acre (113 kg per hectare)! What's not to love about cover crops?

When we were growing mostly vegetables, we would leave at least one-third of our ground available for cover crops. Our favorite routine was to put in a series of cover crops that included a legume for nitrogen fixation every second or third year after growing vegetables. We would start with a winter rye–vetch mixture in September or early October, after the veggies were harvested, scratching the soil with a shallow-tine harrow (or even a piece of chain-link fence bent into small spikes with cinder blocks on it) pulled by our horses. The light raking would also pile up the veggie and weed residues, which we could gather and compost. This left bare soil exposed for us to broadcast the cover crop seed. Another pass with the harrow would guarantee good seed-to-soil contact for cover crop germination.

We would get fall growth from the rye and vetch with this method, but the real growth would occur during the following spring. The abundant spring growth would generally need to be plowed under to establish a seed bed, but the best technique was to graze this overwintered bonanza with sheep, chickens, rabbits, or even pigs to rototill with their snouts (see figure 2.6 on page 27). We loved taking advantage of the free animal forage as well as the spread-your-own-manure ability of the livestock. If this grazing was finished early in the spring, we would follow with a cover crop of peas and oats, which grow fast in cool weather and could provide a cover and grazing supply for later in the summer. Rabbits especially relished grazing peas and oats through the bottom wires of the rabbit tractors we made. If the grazing was delayed past mid-June, the following cover crop would be buckwheat. Buckwheat germinates in warm weather and flowers in about forty-five days, so a July planting would give us August blooms for the bees, and the quick and thick growth smothered any weed seedlings. Yes, we would graze buckwheat too!

To get back to vegetables for the next spring, we would plant oats in the fall. The oats would winter kill but provide ground cover. These

FIGURE 4.5. Rabbits grazing on a pea and oat cover crop. One doe and her litter reside in each rabbit tractor.

were easy to plant into the next spring. All the grazing and cover-cropping fed the soil and set us up for fantastic vegetable yields for the next couple of years until we repeated the cycle. While we don't have much exposed soil these days, because we're using no-till techniques with mostly perennial plants, occasionally we'll create a bare spot (by using landscape fabric to kill the sod) that needs a cover crop. When we first make a hügelkultur, we often start with a cover crop of buckwheat in the first year.

FIGURE 5.1. Flooding from Tropical Storm Irene, August 2011, filled our barns, backyard, field 6, and other areas of the farm.

CHAPTER 5

adapting to climate change

Many factors during our middle years of farming led us to revisit our farming goals with ecological and economic resilience, biodiversity, and regenerative perennial fruit agriculture in mind. We've mentioned a couple, such as changing markets, our aging bodies, and competition from other local organic vegetable producers. But the urgency of annual vegetable production also added a fair amount of stress to our summer. If we didn't get our vegetables seeded and weeded on time, yields would suffer, and we might not have enough shares for our CSA. While perennials still require time management around pruning and weeding, these can be done during larger windows of time (and perennials don't need to be replanted every year!). The loss of pollinators across the region was another factor. We knew we needed to be more proactive in that area by providing more habitat and pollen and nectar resources.

By the early 2000s, climate change had moved beyond theories and computer simulations. It was happening. It was measurable, and the effects were becoming deadlier. We didn't know what it really meant for our region and our farm, but likely scenarios for the coming years included more extreme weather events (floods, droughts), extreme temperature variations in the winter caused by polar vortices and warm spells, and an overall increase in precipitation. For many reasons, biodiversity loss was increasing around the planet. It was clear that we needed to make an even greater push toward a resilient regenerative farm. When John was once asked what he would do if he had only six months to live, he replied without hesitation, "Plant faster!" We now have that same sense of urgency, and we are planting faster.

Flooding

We've always had regular seasonal flooding in our field 6, which is part of the Lamoille River floodplain. Our first big flood came in 1995, and was the most difficult because we were caught completely off guard. We'd lived in the place only a few years, and experienced a mini drought that summer. When family visited in July, the Lamoille River was so low they were able to walk across it. (That turned out to be a bad idea because when they got out of the river, several of the kids had leeches on their legs.) Our spring that fed the house was low, and with so many people, we had to conserve water. Our soil was also parched and dry.

Then in August, rain started one Friday afternoon up near Hardwick and Greensboro, the source of the Lamoille River. It rained and rained and rained there for three days straight, yet we didn't get much precipitation farther down the river in Jeffersonville. All that rain at the river's source filled the small streams, which drained into the Lamoille. It filled the river's banks, then overfilled the banks and into the floodplain. It rolled down the eighty-five miles of the river like a gigantic slow wave, rising and falling as it passed. When it rose, it flooded towns and land near its banks, then receded as the wave moved to the next place downstream.

It was a beautiful sunny August day in 1995 when the river started rising in Jeffersonville. John was at the neighbor's, helping pump out their basement and moving their stored items. Nancy was in the backyard with Connor and Nolan (then eight and four years old). The back driveway, backyard of the house, the farm stand, and barn, all sit lower than the stream bank, so when the water overtopped it, these lower areas started filling up like a bathtub. Nancy called John to come home and by the time he returned, the water was a couple feet deep in the back driveway and the barn was filling fast.

adapting to climate change

Brooding chickens and turkeys got wet before we could get to them. We were wading in the deepening water, moving equipment, tools, and other stored items up on shelves and moving animals to higher ground. Within another hour, the backyard, barn, driveway, and farm stand had over three feet of water, and the water had reached our vegetable garden on higher ground. The boys stood on the porch and watched us frantically trying to save our stuff.

When we saw a mouse floating on a board on the flooded driveway, we thought of scenes from the children's books we'd read to the kids, like *Stuart Little* and *Redwall*. It wasn't only our equipment or infrastructure or farm animals that were affected; the plants and animals both large and small that played a role in the farm ecosystem could not escape this dramatic event.

Within a day the wave had receded, leaving behind damaged coolers and equipment, dead baby chicks and turkeys, soggy silty Reemay row covers, boxes, fencing, and hundreds of items that had been stored on the floor or lower shelves. What a mess to clean up! But we were safe and the house on high ground remained untouched by floodwater.

That was the first flood, the unknown, the disbelief at how quickly water could flood the historic barn and backyard. How truly mighty and destructive the river can be. When the five-hundred-year flood hit in the spring of 2011, we may not have been ready, but since we'd gone through a flood before, it didn't feel as bad as the first time. It took the second five-hundred-year flood from Irene, like a clunk on the head with a two-by-four, to get us to wake up. While the April flood made a mess in the barn and damaged equipment, it happened before the farming season began. We didn't lose any crops. August was a different matter. Our plantings of potatoes and corn in field 6 were ruined. We were still running our vegetable CSA out of the farm stand at that time, so this loss was a big deal. Yet the flood-tolerant perennial shrubs we'd planted, including perennial berry bushes and conservation shrubs we'd planted near the driveway and the reclaimed lawn, didn't suffer at all. Even newly planted berry bushes that were partially flooded seemed to take it well. In central and southern Vermont, many farms lost not only crops but buildings, animals, and even acreage as rivers changed their course. Compared with those farmers, we felt lucky but were concerned about future floods and droughts.

Those events drove climate change home for us. In other words, it was getting wetter, drier, warmer, and even colder at different times of the year than it had in the past. During the height of the Irene flood, we decided the fate of field 6. No longer would we plant that field with annual vegetables. We were going to plant it with elderberries and aronia. Both are tolerant of heavy soil and occasional flooding.

Riparian Zones and Grassy Waterways

Over the last several decades, the Lamoille Valley has been experiencing an increase in the amount of annual rainfall by about nine inches (23 cm) per year (from 40.5″ to almost 50″ / 1–1.27 m annually today). National Oceanic and Atmospheric Administration (NOAA) data also show more extreme precipitation events in our area. Land development has increased the amount of impermeable surfaces such as roads, roofs, and parking lots, resulting in more runoff and thus more flooding downstream. Although it's been wetter overall, it doesn't mean that the rainfall is uniform. Individual storms are often of greater intensity.

Occasionally, and more frequently in recent years, we get a heavy rain that sends streams of water shooting down the hills and through the back fields of the farm. Our small seasonal stream turns into a raging river overflowing its banks and sending water over field 6. A flash flood like that can erode bare soil, which it occasionally did when it was a vegetable field. After planting it with rows of aronia and

FIGURE 5.2. A flash flood is shown in field 6 before we had planted it with perennials. We used to plow this field.

elderberry, bare, potentially erodible soils are a thing of the past. Between the fruit rows, a mix of clovers, grasses, and forbs acts as a grassy waterway, protecting the soil from erosion even during fast water. We also use these grassy alleys as a fertility source for the fruit when we "mow and blow" the grass into the row.

Restoring the riparian zone of our seasonal stream has been an ongoing process. Previously, the stream was regularly dug out and straightened to remove water off the land as quickly as possible. One of our early mistakes was to listen to the locals and have it cleaned out. The water ran brown like chocolate milk through the stream and into the river for a good part of that summer. Since then, we vowed not to scrape it again. By slowing down the water, the land can collect nutrients and create biodiverse habitats. We wanted the water zones to be thick with shrubs and trees, which protect the stream from erosion, provide shade that maintains cooler water temperatures (good for aquatic animals including fish), reduce overland flow of sediments, take up nutrients and other pollutants, and provide diverse niches for plants, insects, birds, and other wildlife. Wooded riparian zones also provide wildlife corridors that help extend the range and the movement of various birds and mammals. Plowing and planting crops to the edges of rivers and stream as well as mowing or grazing, a common practice on many farms in Vermont and elsewhere, undoes these beneficial aspects. In 1992 our farm was no different (see figure 1.2 on page 8).

While there were other reasons to restore the riparian zones on the farm (providing firewood, woodchips, space for wildlife, and screening us from the previous owner's watchful eyes), it was also critical for helping to protect the stream from bank erosion during these more frequent high-intensity storms.

Maintaining grassy waterways in our back fields for spring flooding has also been essential to protect land from erosion. In these high-flow areas, trees and shrubs could create high water turbulence around their trunks, leading to scouring and soil erosion. When the same area is left as grass, the water flows uniformly; the thick grassy sod protects the soil, and grass stems help slow the water down. That's our goal, after all, to slow the water and allow it to soak in.

Wet, Heavy Soils

Diversifying crops across a range of locations on the farm to match plants with their preferred soil types has been a way to increase biodiversity, and thus ecological and economic resilience, by optimizing production. We've planted elderberry plants not only in areas like field 6 that experience seasonal flooding, but also in areas where the soils are heavy and contain more silt and clay. In these areas, we've also planted silver maples and several varieties of willow shrubs for biomass to make woodchips. These plants bloom in the early spring and serve as sources of pollen and nectar for bees. We often cut the pussy willow stems in March and bring them inside to force the flower buds. Then we can sell them as spring-cut stems at the winter farmers market. We also propagate willow cuttings for our nursery and use stems for making baskets and wreaths. These are examples of stacking functions for multiple benefits.

In areas where the soil is heavy, we've created raised beds. In field 3, for example, we plowed furrows and mounded up the soil between the furrows. We then planted black currants, gooseberries, and plum and pear rootstock on the mounds. Even though the furrows are only about a foot or two lower as compared to the top of the mounds, this has been enough to help drain the soil and allow the plants to thrive.

Regenerative No-Till Soil Practices

After our first few years of vegetable growing, we noticed we were doing harm to the soil structure. Too much rototilling and cultivation were causing crusting and cracking of the soil surface. The tilth was being harmed by overworking and allowing the organic matter oxidation by mixing air into the soil. After we realized that, we tried to minimize working the soil, but we were still regularly disrupting the soil ecosystem with tillage practices.

Back around 2008, while we were still growing vegetables, we experimented with no-till methods for vegetables. One technique involved rolling and crimping a winter rye cover crop in the spring when the rye plants were beginning to flower. Rolling and crimping involves rolling down the plant and then crimping the stem. If we just mowed it, it would regrow and produce multiple stems (called tillering). Crimping kills the rye and thus prevents tillering. We could then hand-transplant pumpkin starts into the thick dead rye mulch. The crimping worked pretty well. As mentioned in chapter 3, the additional benefit was that the pumpkins stayed cleaner as they grew by resting on the residue.

When we moved to growing mostly perennials, we shifted to no-till practices, and currently, that is all we do. In other words, we don't turn over and mix the soil. Not mixing the soil also encourages an expansive growth of good fungal hyphae as they work their way around plant roots to provide nutrients in exchange for sugary root exudates. No-till also helps keeps moisture in the soil. We occasionally think back with a bit of guilt at all the plowing, rototilling, and oxidizing of the soil we did in the early years. You wouldn't have wanted to be an earthworm or soil microarthropod (insect, spider, pill bug, or pseudoscorpion) during those times of slicing and dicing.

When growing perennials, we find planting and maintenance in a no-till system relatively easy. Dig a hole, plant the fruit tree or berry bush, top-dress with compost, water in, and use either a sheet mulching technique with cardboard and woodchips or woven landscape fabric to keep the grass and weeds from competing with the plant for nutrients and water. The cardboard breaks down in a season and adds to the soil carbon. The landscape fabric is applied in two strips with the plants in the center so that the fabric can be rolled up in the fall and soil amendments can be added either then or in the spring.

FIGURE 5.3. This practice of using tillage for weed control in the early years resulted in damaged soil structure and a disruption of the soil ecosystem.

FIGURE 5.4. Sarah adds compost to elderberries after opening the landscape fabric. She will close it afterwards.

We've been experimenting recently with no-till annuals such as pumpkins, sunflowers, and high-cannabidiol (CBD) hemp in the alleys between apple trees in our Knoll Orchard. This is a way to take advantage of the open sunny space in the orchard alleys while waiting for the trees to mature. These annual crops were all transplanted by digging holes into the sod and using strips of woven landscape fabric to keep the weeds from encroaching. It's worked well. Next season we should be able to direct-seed beans, beets, and sweet corn into the soil that was under the fabric in those alleys. We will make shallow (½-inch / 1.25 cm deep) furrows and cover the seeds in the row with compost. Can't wait to try that out!

Mulching in a No-Till System

Sheet mulching with cardboard and woodchips, as well as the use of woven landscape fabric, has been essential for growing our perennial fruit, nursery stock, and other plants. We've found it the best way to manage the weed competition.

For cardboard, we stick with food-grade brown cardboard that doesn't feature any colored inks, so as to avoid introducing any dyes or unknown chemicals (the black type on the boxes is printed with

FIGURE 5.5. Cardboard can provide weed management with no-till planting.

FIGURE 5.6. Ramial woodchips in the cart are ready to use as mulch.

soy-based degradable ink) into our soils. We collect our cardboard from a local grocery store dumpster, from the inside of which we often find ourselves waving to friends and acquaintances. Dumpster diving is good for keeping our egos in check and giving our neighbors something to chuckle about.

To sheet mulch, we lay the cardboard around the base of our plants, then top it with woodchips to keep it in place. The cardboard prevents the weeds and grass from taking over, keeps moisture in, and breaks down over the season, thus adding more organic matter to the soil. If done a year in advance, this is also a great way to start a new annual no-till planting bed.

One of our favorite types of mulch, either on top of cardboard or by itself, is ramial woodchips. Made from branches and saplings less than four inches in diameter, these woodchips have a carbon-to-nitrogen ratio of around 30 or 40:1. This is because of the higher ratio of actively growing nitrogen-rich cambium under the bark compared to the carbon-rich lignins in the interior of the branch. The C:N ratio of 30 or 40:1 is the same ratio we aim for when starting a compost pile. Piles of ramial woodchips will degrade over time. Our favorite species to chip are silver maples, box elder, and willow. These grow fast and can be coppiced every few years. "Coppicing" means cutting certain

FIGURE 5.7. No-till elderberries were planted directly into sod. The woven landscape fabric is used as "mulch." Riparian zone trees can be seen in the background.

trees that will resprout for later harvesting. Silver maples, willows, and box elders also grow readily in our riparian zone and other wet areas on the farm, so we can take advantage of marginal soils to grow fertility (woody shrubs that we later chip into mulch) and import it to our fruit-producing areas. This mulch helps promote beneficial fungi with their white mycelia that we often see growing through the piles.

Because we are such organic-material scavengers, we also take generic woodchips from landscapers and others looking for places to dispose of them. If John sees a woodchip truck, he'll pull in next to the truck and talk the guys into dumping the next load at the farm. Once when he saw the truck at a convenience store, he went into the store and tried to figure out who was driving. There were about half a dozen likely candidates in the store. John looked them over carefully, used his intuition ("gift"), and found the driver. Or maybe it was the woodchips on the guy's pants. We like to give drivers a little cash or gifts of jam or honey for dropping off their mulch.

Growing our own mulch is perhaps our most elegant practice. The alleyways of our orchards and other areas between plants are places where we can grow biomass and fertility to be cut and placed where we want it. One simple example is to mow and blow the alleyway's material into the crop-producing row using a walk-behind mower. Other mulches we grow are comfrey, sunchokes, and rhubarb leaves (after harvesting the stems). These are "chopped and dropped" for mulch around crop plants. Pumpkins planted near our trees also provide a living mulch.

When our nitrogen-fixing trees such as black locust, honey locust, speckled alder, and Siberian pea shrub grow a little more, we also plan a chop and drop system for the branches. Besides mulch, the repurposed plant debris provides nutrients to the plants when it breaks down. We also want to experiment with creating mulch from cattails and other aquatic plants that we can grow in our wetter areas.

Woven landscape fabric has become indispensable for us as our substitute for herbicides. We roll out long three-foot-wide strips along each side of fruit rows such as elderberry and blueberry and use five-inch wire staples to keep these in place. The strips kill the grass and weeds around the plant by occultation (a fancy way of saying "lack of light"). In the fall, we roll up the strips to keep the vole population down. Otherwise, these little rodents would party under the fabric shelter through the winter, which is when they often eat the bark of trees and bushes, causing a lot of damage. They can completely girdle the bark of a tree or bush and kill it. We have also learned the hard way

adapting to climate change

not to put woodchips on top of woven landscape fabric. This encourages quack grass and other weeds to grow through the fabric, which entails backbreaking effort to pick up the fabric.

Hügelkultur Practices

Another beneficial (and fun) landscape-enhancement strategy is our hügelkultur practice. In permaculture circles, hügelkultur—a German word meaning "mound culture"—is a technique used to provide a reservoir for moisture and a slow release of nutrients from wood and prunings as they decay. We make piles of woody prunings from our fruit trees and berry bushes, old pallets and waste lumber—anything organic, really. There might even be old cotton jeans and T-shirts in these piles. When deemed tall enough, usually three to four feet high, we cover the piles with a few inches of soil and compost.

Currently we have about eight hügelkulturs in various stages on the farm. It's a good way to create fertile raised beds as the decomposing materials slowly release nutrients and hold water. After covering

FIGURE 5.8. The start of a hügelkultur near field 3 shows the prunings and wood that make up its interior. These will eventually be covered with soil and planted with a cover crop, vegetables, or flowers.

the mounds with soil, we plant them with annual cover crops that also provide nectar and pollen for the bees the first year or two. Later, we might plant flowers (both perennials and annuals), herbs, squash, and vegetables. After a few years, the piles settle to about half their original height and become a nice permanent raised bed. It is a vast improvement over burning prunings and sending that precious organic matter into the sky as greenhouse gases. We call our piles "bumblekultur" because an added advantage is that they provide nooks and crannies for overwintering bumble bee queens and nesting nooks for other critters.

Dry-Weather Watering

The problem with climate change is that we don't really know what it means at a given location or time. Models have predicted wetter weather in New England, but as this area also continues to warm, that could change. The summer of 2018 was the hottest and driest that we've had in the twenty-six summers we've lived here, with severe drought conditions in our area. We can't count on cooler, well-watered summers anymore.

Even though we're usually blessed with a good amount and even distribution of precipitation in the summer, we typically do have drier weather in August. The groundwater table gets low, which means that the level of water in our spring gets low too. The spring provides household water and water to an outdoor frost-free hydrant. During low-flow episodes, we are mindful of water use. Running drip irrigation for vegetables in the early years was often done at night, when household demands were low.

In recent years and especially during the summer of 2018, watering nursery plants in our on-farm nursery solely with water from our spring has not been possible. We have enough pressure to fill our household cistern if we are conservative with water use, but not enough pressure to use the hoses for watering. This has prompted us to increase our water-harvesting and storage capabilities. In 2018, we added a third 275-gallon (1,040 L) tank to our existing two that collected water from the extensive house

FIGURE 5.9. Water collection and storage tanks can also serve as a support for Concord grapes.

roofs during spring and summer. We also moved an 800-gallon (3,025 L) tank that we'd used for watering livestock in the back pasture (when we had livestock) to a new spot near the nursery to collect rainwater from the barn roof. Two smaller 50-gallon (190 L) tanks also collect water from the barn roofs and are used for watering horses and plants. Even a quarter to half an inch of rain will partially fill the tanks and provide a couple of weeks' worth of water for nursery plants. When the spring is running sufficiently, these storage tanks are still beneficial because of their strategic locations. They are elevated and provide a higher water pressure, which makes it easier and faster to water nursery stock and propagation beds. Storage from roof runoff was a practice we saw firsthand while working with farmers in Haiti, the Dominican Republic, and Africa. This is a good example of how horizontal farmer-to-farmer knowledge across geographical areas can be an important part of addressing climate change challenges.

Repurposing Unheated Hoophouses

From 2004 to 2006, like many CSA and farmers market veggie growers in Vermont, we put in several thousand-square-foot (93 m²) unheated plastic hoophouses (21' × 48' / 6.5 × 14.5 m) for growing tomatoes and other vegetables. We added two more in 2011, bringing our total to five houses. Funding for three of these came from a USDA Natural Resource Conservation Service program. Tomato plants like hot, dry conditions, and our typical moist, humid, and often cool summers in Vermont can delay their growth and invite diseases. Hoophouses provide a protected environment. Cherry tomatoes were one of our favorite vegetables to grow, and we found good local markets at two of the small groceries stores nearby. We used cover crop rotations during our tomato-growing years, and every few years we took the plastic off one house during the growing season to let the rainfall recharge the soil.

The hoophouses increased our resilience in the face of extreme weather because keeping rainfall off our plants reduced occurrences of leaf and fruit disease. We could shift the temperatures inside by rolling their sides up or down, which extended harvests earlier and later into the season while protecting the plants from frosts in the spring and fall.

By 2009, we decided to plant a few rows of fall-bearing raspberries in one of our houses. Our favorite varieties are Polana and Joan J, a thornless variety. We love fresh raspberries, which are hard to find locally, especially organic raspberries (conventional berries are typically sprayed with fungicides and other pesticides). That first raspberry row in the hoophouse did so well that we filled another hoophouse the next year. We continued adding more rows in other hoophouses after that. Hoophouse raspberries have been a great crop for us, commanding a good price for fresh berries at the farmers markets, as jam, as syrups for snow cones, and wholesale to small juice and beverage companies.

Raspberries, like tomatoes and peppers, prefer hot and dry growing conditions. Hoophouses create the conditions they like. Planting fall-bearing raspberries under cover can also hurry them along so they start their annual bearing a bit sooner than they would if planted outside. If we have an early September frost, we can also close up the house. We often pick well into October. But the best thing about growing raspberries under cover is the protection from rain and moisture, which encourage moldy berries.

adapting to climate change

There isn't anything much better than going into the raspberry hoophouse on a gorgeous summer day and picking (and eating) raspberries. They are warm and plump and sweet and sour at the same time. When we're picking them for fresh market, we try to eat only the misshapen ones or the ones that a Japanese beetle sucked a hole into; but when we're picking them for the freezer, we don't have to limit ourselves to the seconds. We can find the biggest perfectly ripe berries and eat them up.

Many of our raspberry canes are so tall that the berries can be reached only by gently tipping the canes down. We move carefully through the canes so as not to break the tips where most of the flowers and new berries are. But there are other reasons to move slowly and mindfully through the canes. The thorns, for one. Even the leaves are scratchy. There are a lot of bees and an occasional wasp in there too; the bumble bees to drink nectar and collect pollen, and the wasps look for damaged fruit to suck on or nectar from flowers for energy. Wasps

FIGURE 5.10. Nancy takes picking raspberries very seriously.

FIGURE 5.11. Dwarf apple trees bloom in a hoophouse.

might also be looking for soft-bodied insects to take home to feed their babies. Moving a hand too quickly to pick a berry might result in a needle-like sting from a surprised wasp, but they'll get out of the way if you move slowly. Other welcome insect predators within the raspberry canes include earwigs, daddy long-legs, and other spiders.

The bumble bees are the most abundant of the big insects in the raspberry house, and they aren't prone to stinging. They never threaten or even seem to care when humans are around. We each wander the rows doing our own thing. They're fun to watch. Maybe they feel the same way about us. Amazingly good at pollinating, they move from flower to flower, vibrating them at the perfect pitch to release pollen, a technique often referred to as buzz pollination. Each raspberry is actually composed of a bunch of small ball fruits, called drupelets, fused together into a thimble shape. Each one of those drupelets needs to be pollinated for a fruit, with its seed, to grow. That's why raspberries are so seedy; there may be a hundred or more drupelets (and seeds) per berry. When you see a misshapen berry, it might mean that not all the flowers were pollinated.

In 2012, we saw the first evidence of a new pest in our raspberries called spotted wing drosophila (SWD). SWD is a pest of soft fruits that has made growing no-spray berries a whole lot harder, but still doable at our farm. (We'll talk about how we deal with SWD and other "pests" in chapter 10.)

These days in Vermont, growing tomatoes and raspberries in hoophouses is fairly common. What's less common is growing dwarf apple trees and apricot, cherry, plum, and peach trees in hoophouses. A few years ago, we phased out of the cherry tomato operation. It was profitable, but the urgency in getting the tomatoes picked and filling orders every couple of days became a bit of a distraction from our fruit focus. Besides, we had grown tomatoes for a long time and were ready to learn new ideas. Retiring the tomatoes opened up space in the hoophouses and in our minds.

Why not plant fruit trees in the hoophouses as a resilience strategy? Rain and moisture can result in many different disease and pest problems, so keeping the trees undercover makes a lot of sense if you have the space. We've found growing high-density dwarf apple trees undercover has been a great way for producing high-quality apples for fresh eating with zero pesticides.

Climate change in Vermont has thus far resulted in warmer weather overall. It's also resulted in more instability, especially in the winter. Increased temperature swings from warm to cold in the winter can be tough on plants, especially with no snow cover. We've seen warmer springs, but we're still susceptible to late May frosts. These conditions can wreak havoc on plum, cherry, and peach crops. Growing them in a hoophouse could be beneficial. We can close the hoophouse and prevent a late frost on fruit or blossoms. We can keep birds and insect pests out with screening and prevent fruit cracking from too much rain near harvest time. We're still early in the experiment for the stone fruit but encouraged by a few tasty plums from young trees this past summer.

While plastic unheated hoophouses have been a great addition for our organic tomato, pepper, ginger, raspberry, and now tree fruit operations, they are not without their own problems. Clearly, there are initial capital costs. We found that by selling cherry tomatoes at our local grocery stores, the hoophouse paid for itself and our labor within one season. Raspberries require a couple years to get established but once they do, they easily pay for the cost of the house and don't need replanting every year. We prune the everbearing varieties down to the ground in the winter, and they regrow in the spring. It is too early in our trials to calculate the economics of growing tree fruit in our hoophouses, but in our case, the hoophouses were already paid for when we started.

Upkeep is also a challenge with these plastic hoophouses. We have experienced plastic and end wall damage from wind and ice storms on several of the houses. Rolling up the plastic to the top and lashing it to the ridge purlin is one way to avoid snow and wind damage in the winter. This also helps with flushing the soils from salts that can build up from the compost. We recommend it every few years. It's not easy to roll them up like that, though, and it can cause a bit of damage to the plastic as well.

It's good to change the plastic every three to four years, as it becomes less transparent over time, which reduces the amount of sunlight that gets through. If you wait for a day without wind, changing the hoophouse's plastic (called reskinning) is not too difficult, but for us it's often one of those chores that gets put off.

We don't do much irrigating or watering in our hoophouses. Again, tomatoes, peppers, and raspberries like it dry, so except when first establishing them, we don't need to water or irrigate. It seems magical that we can produce big raspberries and other fruits without watering. Our hoophouses are only twenty-one feet (6.5 m) wide, so we imagine the runoff from the plastic is brought back to the center of the house by the capillary action of the roots sucking up water. By not

FIGURE 5.12. Honeycrisp apples (unsprayed) grow well in our hoophouse.

FIGURE 5.13. Stone fruit like these apricots can do well in a protected hoophouse. In this case, accompanied by rows of tomatoes, peppers, and raspberries.

irrigating the surface, the plants are also trained to send their roots deep. Not irrigating is one of the reasons we think our raspberries are so tasty. Their raspberry flavor is concentrated, not watered down. We also thought that was the reason our cherry tomatoes didn't split so much even late in the season. They weren't getting excess water when the fruit was ripening.

We watered the trees we planted in the hoophouses occasionally during their first year after planting, when their roots were getting established. Fruit trees and bushes also need sufficient moisture for fruit set and maturity. For our trellised apples, which are on shallow-rooted dwarfing rootstocks, we've drip-irrigated them in the early parts of the season, especially when rainfall has been sparse. Generally, though, we haven't felt the trees are lacking water. They seem to be able to draw enough water to their roots from the water table or from beyond the hoophouse walls. As with everything else on the farm, we keep an observant eye on them.

CHAPTER 6

agroforestry in action

When we decided to expand our fruit tree orchards around the farm, we wanted to deliberately incorporate more agroforestry practices into these perennial polyculture orchards. We weren't new to the concept of agroforestry. Establishing hedgerows, growing trees for biomass, and enhancing and maintaining riparian forest zones next to streams are agroforestry examples that we had already been practicing on the farm with great results.

Our riparian zone of trees and shrubs has created bird and pollinator habitat that benefits the farm and gives us a lot of enjoyment. Bees and other pollinators use the early pollen from the willows, silver maples, and other trees. The diversity of pest-eating birds visiting and nesting there is a designed part of our ecological pest management strategy. The economic and ecological benefits are many.

Creating hedgerows as windbreaks is another valuable agroforestry approach that can enhance biodiversity, reduce soil erosion, and protect animals, buildings, and crops. Our dominant winds blow from the west, so we've been planting a north-south-oriented hedgerow that bisects what used to be called our back pasture and is now called the pollinator sanctuary. We grow black locusts, honey locusts, elms, hawthorns, osage orange, elderberries, dogwoods, and black walnut

FIGURE 6.1. A cedar waxwing nests in the riparian zone.

trees as a future windbreak and resource. This mixture of tall trees and shorter shrubs will eventually create a layered canopy, which is beneficial for bird and wildlife diversity as well as blocking wind. The hedgerow will serve as a wildlife corridor for connecting the river to the forested area north of our farm. We want to encourage the movement of higher-level predators such as coyotes and foxes across our farm to help keep deer and rodent populations in check. Birds and insects can also take advantage of this flower, berry, and insect source, while moving through the cover of this ribbon of connectivity.

A Perennial Polyculture Orchard: The Fruit Trees and Guild Plants

In 2012, we started planting two small orchards in our pollinator sanctuary: a one-acre orchard of primarily heirloom apples (Knoll Orchard) and a half-acre pear orchard called Pear Corner. We also have a few rows of commercial and native plums alongside the apples in the Knoll Orchard, and Plum Alley contains another twenty plum trees (see figure 1.4 on page 10). Other crops are mixed into these orchards, but the fruit trees are the foundation. We jokingly call this whole setup the "conventional orchard," as we would love to see these regenerative practices that make economic sense, sequester carbon from the atmosphere, slow down runoff, and increase biodiversity become the new dominant conventional agricultural and land-stewardship model.

For optimal soil health and to reduce erosion, all our crops are grown without disturbing or tilling the soil. The trees are planted twenty feet (6.1 m) apart in rows twenty feet apart. The rows run east-west, allowing the sunlight from the southern aspect to be available for the plants between the trees. Many of the apple trees are on standard rootstock (Antonovka or Bud 118) and should reach at least twenty feet tall, hence the twenty-foot spacing. Likewise, the pear trees will grow to at least twenty-five feet tall on their pear rootstocks (Old Home × Farmingdale, 97 or 333). The primary goal of these orchards is to produce unsprayed fruit for hard apple cider and perry (pear cider) production.

The Knoll Orchard is planted on a silty-loam knoll deposited when the farm was under a glacier-meltwater lake, eleven thousand years ago. The silty soil is not so desirable for apples as is the sandy loam of our Front Lawn Orchard, where we planted twenty-four apple trees years ago. The silty, heavier soil seems to have slowed down the trees' growth a bit in comparison. Many of the Knoll Orchard trees are standard apple trees, which also are slower growing and less precocious for bearing fruit than the semidwarf rootstock trees on our sandy front lawn. The sunny south side of the knoll, however, is ideal for both water and air drainage, which makes it a good site for trees and intercropping. The newer pear orchard is planted in a higher, drier sunny location in the back meadow (Pear Corner).

While the Knoll Orchard started with twenty-four heirloom and cider apples, we've been adding to it ever since, planting different varieties each year. The biodiversity of apple trees allows us to collect scion wood for selling and grafting new trees, and for getting to know, taste, and use different apple varieties in our products. Biodiversity can also be promoted within a species by the genetic diversity of the different varieties and will confer resilience in the form of different flowering times, pest resistance, fruiting cycles, ripening times, and fruit-storage potential. Since standard apple trees can live up to one hundred years, we also see these trees acting as a repository for heirloom apple variety genetics. Hopefully, the Knoll Orchard will remain a legacy long after we're gone.

Another exciting aspect about these orchards has been the opportunity to create supportive guilds by planting a variety of shrubs and perennial flowers within the orchards. These provide benefits to the fruit trees, the farm, and us. We plant the guilds between the fruit trees, within the tree rows. We don't always follow a standard pattern, but a row might look like this: A nitrogen-fixing tree or shrub is planted in the center between the apple or pear trees. On either side of that nitrogen fixer—about seven feet (2.1 m) from the fruit trees—we plant shrubs that bear fruit for the benefit of us or of wildlife. The shrubs that mature more quickly than the trees can take advantage of the space (and light) while the trees mature.

Black currant bushes, for example, take only a year or two before their first harvest, while the apple trees could take ten or more years to bear a fruit crop. We've harvested black currants from the Knoll Orchard for the past two years. The sunny location and open spacing help the berries ripen a bit earlier than the other denser hedgelike plantings in fields 2 and 3. The open spacing also makes for easy picking. Black currants don't mind shade, so even when the apple or pear trees are fully grown, the currants should still be quite happy in the partial shade of the big trees.

The nitrogen-fixing trees fix atmospheric nitrogen and may help provide nitrogen to the fruit trees over the years. When the trees and shrubs are older, we plan to chop and drop their branches to provide mulch and long-term fertility for our fruit trees as the leaves and branches decompose. When plants are cut or pruned, there is a corresponding shedding of roots that also occurs. We hope this provides

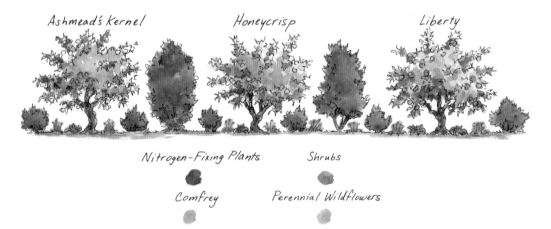

FIGURE 6.2. This illustration shows the planting pattern between apple trees in the Knoll Orchard. *Illustration by Elara Tanguy.*

extra nitrogen from the pruned plant roots to be used by nearby fruit trees. For now, it is hopeful thinking. Quantifying this potential source of nitrogen would be a great area for agroecological research. Bayberry may also provide its own viable crop someday, as the fruits are used in candlemaking and other crafts. Other shrubs, such as different species of dogwood and wolf willow, have value as cuttings and products for the floral industry.

Nonwoody perennials are planted between the shrubs and slightly off center of the row in the southern direction to minimize shading. These include native flowers that provide beneficial insect habitat. A plant like comfrey, when planted near the apple trees, not only provides nectar and pollen for bumble bees, other pollinators, and predatory wasps, but will also be scythed to provide a green mulch around the trees and shrubs. Comfrey has a large taproot, which brings up nutrients from deeper soils and returns them to the surface in the leaf mulch. Most of the fruit tree feeder roots are close to the surface, so there is likely little competition. When the leaf mulch decays, it releases nutrients that can be used by the trees. Comfrey can also be used as a medicinal and dye plant. When we raised chickens and rabbits, we often fed them comfrey leaves as well.

Besides providing alternative agricultural products, the guild plants provide biodiversity in the orchard. Most of these shrubs and perennial flowers are great pollinator plants. They bloom at different times during the growing season, thus providing nectar and pollen resources for bees and butterflies over the course of the season. Most guild plants are also native to Vermont. This means they will be host plants for a whole range of native insects that have coevolved with them, including butterfly and moth larvae, beetles, and more. In turn, these insects provide food for insect predators (which will also eat apple pests), birds, and other wildlife. By creating a thriving biodiverse orchard ecosystem, we are creating a thriving ecological pest management system as well. These perennial polyculture systems can be applied to many scales, from backyard to large farm.

Silvopasture and Alley Cropping

Silvopasture and alley cropping are two types of agroforestry practices that have been designed into our Knoll Orchard layout. Silvopasture integrates trees with grazing animals. In some cases, the trees provide forage or shelter (or both) for the grazing animals. In our case, we're grazing horses between the rows in our polyculture apple orchard while we patiently wait for the trees to mature and start bearing beautiful heirloom apples. The trees are still small enough that they're not providing much shade, but someday they will. Using the horses for management intensive grazing helps prepare these spaces for alley cropping by grazing down the meadow plants and encouraging the growth of clovers and grasses all while the horses spread their own fertilizing manure. We also think about the future potential for incorporating laying hens, ducks, sheep, pigs.... With this system in place, it's tempting to get caught up in livestock dreams again!

"Alley cropping" is a term that describes using the wide lanes between tree rows for growing other crops. This is especially useful while waiting for the trees to mature and begin producing. Once mature, the trees' canopies may provide too much shade for growing most crops. Also, at that time, the alleys will be needed for harvesting and collecting the fruit. For several years now, we've encouraged natural common milkweed stands within the alleys between the fruit tree–guild plant rows and have collected the pods for milkweed seeds and the insulative and hypoallergenic floss. We've sold seeds and pods

FIGURE 6.3. Horses initially grazed the alleyways in our perennial polyculture orchard before planting.

FIGURE 6.4. Pumpkins and sunflowers grew well in an alley of the Knoll Orchard.

to individuals and other businesses. Monarch butterflies appreciate the milkweed stands too! Milkweed, a perennial plant, sends its stalk up later in the spring than other plants, much in the same way that asparagus sends up its stalk. We take advantage of that fact by grazing the horses in the alleys early, before the milkweed emerges, thus reducing milkweed competitors and helping the milkweed grow into thick stands that are easy to harvest.

This past year, we also grew different no-till annual and perennial crops within six of the alleys between the fruit tree–guild plant rows. These crops included pumpkins, hemp, sunflowers, and two perennial crops, sea kale and sunchokes. The horses prepared the site by first grazing it down and encouraging a Dutch white clover resurgence. The clover acts as a nitrogen fixer, and when we covered it with landscape fabric, the decomposing clover fed the crops.

We started the pumpkins and sunflowers in trays in our greenhouse and then planted them within the alleys in June, after the threat of frost passed. We mixed compost into the holes at planting for additional fertility. The addition of the organic matter into this silty soil helped create a more open soil structure for the roots while still providing good water-holding capacity. We watered the plants in well, followed by laying out three-foot-wide (0.9 m) black woven landscape fabric strips on either side of the plants to kill grass and clover. The strips were held down and joined together with five-inch-long (13 cm) metal staples right at the stem of the pumpkin or sunflower. After a few weeks, one quick weeding right around the plant was all that was needed. Both pumpkins and sunflowers each support different types of specialist native bees, squash bees and sunflower bees (respectively), so we liked that aspect of helping their populations. We sold the pumpkins at our neighborhood general store, and used the sunflower heads to marvel at and to support such seed-eating birds as goldfinches and chickadees.

FIGURE 6.5. We planted high-CBD hemp in the alleys of the Knoll Orchard.

This past year, we also used four of the alleys between the apple rows to grow no-till hemp that has high cannabinoid (CBD) oil content. A new company in Vermont provided the hemp plants, which we planted at six-foot (1.8 m) spacing using three-square-foot (0.3 m²) bibs made of black landscape fabric for mulch.

This coming season, when we pick up the landscape fabric from the annual crops, we will use these now vegetation- and weed-free areas for no-till seed crops such as beans, beets, herbs, and flowers and move the pumpkins and sunflowers to new alleys and prepare and plant as we did last year.

Perennial Vegetables

With our commitment to no-till and the ease of caring for perennials versus annuals, we are increasingly interested in perennial vegetables. We have been growing asparagus and rhubarb successfully for over twenty years on other parts of our farm. Our small rhubarb patch generates thousands of dollars of income each year; we sell rhubarb stems wholesale, divide the roots for potted plants in the nursery, and use the juice for our strawberry-rhubarb syrup. Maintenance includes a yearly weeding and one heavy application of composted manure. Rhubarb, with its deep roots, can act much like comfrey in taking up nutrients from deeper in the soil and putting them into their leaves, which can then be used as a nutrient-rich mulch. Asparagus, grown within the rhubarb rows, has been a good early vegetable with early cash flow potential. Rhubarb doesn't close the canopy until we're done picking the asparagus spears. The asparagus ferns grow well above the rhubarb in summer and thus get good sunlight, and honey bees love the bright orange pollen of the fern flowers. We've also sold the airy asparagus ferns to florists.

We also planted sea kale and Jerusalem artichokes, both perennial vegetables, in one of the alleys between the apple trees. The sea kale planting was part of a SARE grant in which we are participating as a grower partner. Sea kale is a perennial plant native to European coastal areas. All parts of the plant are edible, but it's mostly grown for its spring shoots, much like asparagus, and broccoli-like florets. It has a bit of history in the United States, as it was nativized to the Oregon coast as early as 1905 and grown by Thomas Jefferson at Monticello in the early 1800s. A nutritious vegetable, it's high in vitamin C, calcium and magnesium, and B vitamins. (How about EAT MORE SEA KALE as a popular new bumper sticker?)

FIGURE 6.6. Asparagus and rhubarb grow well together.

FIGURE 6.7. Sea kale (*front*) and sunchokes are alley-cropped in the Knoll Orchard. Note the welded wire fencing and screening at base of the trunk in a nearby apple tree that protect it from deer and voles respectively.

Sunchokes, also called Jerusalem artichokes, produce a delicious, low-glycemic-index tuber that we love to roast with a little olive oil. They're also high in iron. They were a staple crop in this region before European settlement, and we are pushing for a resurgence. Sunchokes are easy to grow and have a late-blooming composite flower, which is loved by native pollinators. Because they can be very prolific, we hope to try the upper plant part as a source of mulch for the fruit trees in the future. We are not sure of the market potential for either of the sea kale or sunchokes, but we'll develop that over the years to come. For now, we're eating and enjoying them as we learn.

Establishment and Care of Perennial Plantings

There is more to establishing a biodiverse perennial polyculture system than planting different kinds of fruit trees and bushes, but that's an important first step. Just do it! You also need to follow up with care and maintenance. At present, we grow more than thirty different kinds of edible fruits in a variety of locations on the farm. One of the benefits of growing a variety of different fruits is that they ripen at different times of the season. This spreads out the harvest workload and the cash flow. While there are many berries that ripen in July (currants, gooseberries, cherries, and blueberries), other fruits ripen as early as June (rhubarb and honeyberries) or as late as August, September, and October (plums, aronia, fall-bearing raspberries, elderberry, apples, and pears). This way the picking workload occurs over four to five months rather than only one month. Besides harvesting during the season, we're also busy with weeding, feeding, mulching, and mowing (or scything) around our crops. In the late winter, the workload shifts to pruning.

We generally plant in rows because rows do make it easier to manage, pick, net, and lay out landscape fabric for weed control, but we tend to vary the fruit rows within a given field. This keeps the juxtaposition of crop biodiversity in all our fields high. Given the soil and water variability and other factors that change from one part of our small farm to the other, we've kept our fields relatively small and segregated, but in the long run, this has been better for ecological diversity, and hasn't been overly problematic in terms of management of the crops.

As with the alley cropping, we have generally established our plantings by first intensively grazing the area we want to plant. If that isn't possible, mowing is the first step. After that, we lay out the rows by measuring and staking, and using twine attached to two stakes to keep the row straight. We dig holes with hand tools and plant the bushes or trees at the desired spacing. For fruit trees and berry bushes, we add compost or aged manure as a surface application, and water in the plants. Sheet mulching with cardboard or using landscape fabric helps keep the weeds away from the plant. In especially dry years (like the summer of 2018), we occasionally have to water multiple times. Generally, though, we don't need to irrigate, as it rains often enough, the soils have good organic matter content (water-holding capacity), and the mulch helps keep in the soil moisture.

Fruit trees, especially if they are small, need protection from deer. Deer have recently become both more numerous and bolder on our farm than in years past. We see less hunting, as more of our neighbors are posting their land against it. To exclude the deer from the trees, then, our standard deer fencing method is to use five-foot welded wire fencing cut into ten-foot (3 m) lengths (see figure 6.7 on page 104). We bend the fencing into a cylinder that encircles the young tree. We use a metal T-post with tabs and attach the fencing to the frame. This keeps the fencing from moving around and possibly damaging the tree. A noteworthy tip is to close the fencing cylinder with a few of the exposed metal wires to make it easier to open the fence later. The first year we used these, we made the mistake of closing every exposed wire to the other side. It secured the cylinder but was a bear to undo later.

We take the fence off at various times for weeding, mulching, and pruning, but otherwise, it remains around the tree until the tree is tall and sturdy enough to withstand deer browsing. Even then, we still need to watch the trees. One time a buck rubbed against a newly unfenced tree and damaged its bark and cambium layer. We immediately put the fence back on; otherwise, the buck would have returned and done further damage.

Many shrubs and berry bushes are also susceptible to deer browsing. This is especially true for the ones planted far from the house in the pollinator sanctuary, such as our hundreds of elderberry plants (a deer favorite). We generally don't use deer protection on these, expecting and accepting that deer will do a bit of damage. In most cases, the deer don't kill the plants—although they can slow their growth if they keep browsing. Because we have hundreds of them, there's a dilution effect; the deer tend to nibble and graze here and there rather than eat the plants all the way to ground. For people with only a few bushes, however, it might be worth protecting these from deer because they could kill the plant outright, whereas because we have so many, it would be cost prohibitive to fence them all.

By the fall of the year that trees are planted, it's important to protect them from possible vole damage during winter. Voles are small rodents that live in meadows and fields. Trees are a substantial investment in capital and labor, and voles can break your heart by killing a newly planted tree. They especially like living under landscape fabric or thick, tall grassy areas where they are protected from such predators as foxes, hawks, or our dog, Scout. In the summer voles have plenty of things to eat, but in the winter and early spring before other plants have greened up, they are looking for food. The cambium layer inside the outer bark on trees and shrubs provides a nutritious meal. The voles can easily girdle the bark all around the tree. Once this happens, the tree will die, as without contiguous cambium, the top part of the tree becomes disconnected from the roots. A dead tree may still leaf out in the spring thanks to stored energy, but don't let that fool you; it is dead. The roots may still be alive and send out growing shoots, but the rootstock of a grafted tree will not grow the delicious fruit we want.

For vole protection, we've used hardware cloth, aluminum screening, and plastic wrap guards on our trees, all with good success. Any protective vole guard should be at least eighteen inches (46 cm) tall and rest on the ground surface or be buried an inch belowground. Both aluminum screening and hardware cloth can be left on year-round, provided it's wide enough to allow for the growth of the trunk. A plastic wrap should be removed in the early summer—as it can create an attractive moist place for fungal pathogens and borers during the growing season—and then be put back on in late fall. Voles can also girdle bushes and shrubs, but because these aren't grafted and have multiple stems, they are better able to bounce back.

Scything around the fruit trees and guild plants is our way to keep the grasses and other weeds from competing with trees, shrubs, and flowers for water, sunlight, and nutrients. A scythe is a tool consisting of a long wooden shaft (the snath) with handles, attached to a razor-sharp sickle-shaped blade. It can cover a lot of ground fast, though we love scything for its own sake! Cutting grass and weeds

FIGURE 6.8. John is scything in the young Knoll Orchard. *Photograph courtesy of Alisha Utter.*

with a sharp straight-snath scythe is meditative and satisfying work. In addition, the cut plants not only provide a surface mulch, but their deep roots will shed themselves into the biologically active zone and feed the deep-soil microorganisms. Keeping these grasses and forbs down in the winter also keeps the vole populations away from the trees and shrubs.

CHAPTER 7

our fruit and nut trees

We've always had a few fruit trees on the farm, but since 2008, we've been making up for lost time and planting a lot more. The adage, "The best time to plant a tree was twenty years ago, the second-best time is today," captures our thinking these days. If we'd tried to plant the whole orchard all at once, the trees (and fencing) and guild plants could have meant a fair amount of up-front costs for plants and labor. Our strategy has been to add trees and other plants over time, spreading out these costs. Once we plant and fence the trees and plant the shrubs, a little love and labor and a lot of patience are what's needed. Our Knoll Orchard started with only twenty-four apple trees. It now contains hundreds of trees, shrubs, and perennial plants. And we're not done yet!

If people are willing to learn as they go, we encourage them to get started and not wait until they become experts before planting fruit trees. While berry bushes take only a year or two before they start bearing, fruit and nut trees typically can take three to five years, depending on the species, rootstock, and environmental conditions. Our Front Lawn Orchard (twenty-four trees), for example, took about five years to start producing bountiful apple crops. It also takes about five years after planting to start enjoying crops of native plums,

FIGURE 7.1. The young Front Lawn Orchard in bloom.

cherries (if you can beat the frost and birds), and hazelbert nuts (if you can beat the chipmunks and squirrels). Fruit trees can be generous and giving, but they can also break your heart. They're great teachers.

The waiting period between planting and the first full harvest gives growers a little time to educate themselves about pruning and pest management, which are usually needed by year three. People are often more motivated to learn about something when it becomes urgent. At least, that's true for us. We call it "just in time" learning. Reading about pruning, pest management, and other details can be fascinating when a tree is standing in the yard, needing to be taken care of.

While you don't need to be an expert, it is important to know at least the basics before planting, because many fruits (apples, pears, and many plums, for example) require two different varieties for cross-pollination, which often means you need to plant a minimum of two trees. Siting is important as well. In general, all fruit trees like full sun and well-drained soils. We've lost a few trees to sites that seemed well drained, only to realize they became too wet in the spring. Finding a good site is important.

A few nursery customers have mentioned having a beautiful big cherry or other self-pollinating tree that has never given them any

FIGURE 7.2. Liberty apples grow in our Front Lawn Orchard. "If we had an apple for every time someone said you can't grow apples without sprays, we'd have this many apples."

fruit. While there could be a variety of reasons for that, it often turns out that the beautiful tree is in full shade or has never been pruned. In those cases, it's probably the lack of light causing the problem because sunlight is required to initiate fruit buds. One of the reasons pruning is important is because it opens up the tree canopy so the lower branches aren't shaded. Otherwise, fruit production will be limited to the outer periphery of the tree or could stop altogether. Pruning is also important for disease management, as it increases airflow.

Rich, well-drained soil is another basic requirement. The sandy loam soil in what is now the Front Lawn Orchard was just about perfect. The only slight problem occurred in the first year or two of planting, when the trees were getting established. We needed to check the soil moisture during the summer and water as needed because we didn't want those young roots to dry out. After those first two years, though, the mulch, shading, and greater number of roots seeking moisture were adequate for good growth.

The soil in the Knoll Orchard, as opposed to our sandy front yard, is not quite as good, because of its high silt content. Water moves slowly through the small pore spaces, thus slowing drainage. Planting on the slope rather than the bottom of the hill has helped. Besides

FIGURE 7.3. One year after planting a few plum trees in the hoophouse, we harvested a handful of Alderman plums. This one really was as tasty as it looked.

water drainage, planting on slopes also helps with air drainage, which may help prevent flower damage from spring frosts. Because cold air is denser than warm air, it moves toward the bottom of the hill much the way water flows. While the silty heavier soils of the Knoll Orchard may not be the most optimal soil for fruit trees, apple trees tend to be pretty hardy and are able to grow in a variety of soil types.

Climate change in our area has resulted in warmer-than-normal weather in April and even March, which can cause trees to break dormancy earlier than they did in the past. Although our spring weather tends to be warmer, it is nothing to celebrate, because the possibility of May frosts still exists. If the trees break dormancy early, tender tissues may be exposed to spring frost events. This is one of our reasons for trying to grow stone fruits in our repurposed hoophouses. We keep them open to the ambient temperature all winter and spring, but if a spring frost is predicted during or after bloom, we can close up the hoophouse and hopefully prevent frost damage. The hoophouses also keep these trees in the kind of aboveground dry climate conditions with adequate belowground moisture that they love.

Rootstocks, Scion Wood, and Grafting Basics

Fruit trees can be grown from seeds, and many of the great varieties we enjoy today originated from seedling trees. However, if you plant the seed of an Empire variety apple, for example, you will most likely get a completely different apple from the new tree. Each seed harbors its own random combination of the DNA that came from its mother (the Empire tree that made the fruit) and from its father (the tree whose pollen was used to pollinate the fruit). This genetic combination from apple sexual reproduction is expressed in unpredictable ways, most likely resulting in an apple not as desirable as its mother. If you want another exact replica of the Empire, it needs to be propagated asexually, by cloning. We do this by grafting scion wood from the original tree onto a rootstock or a branch of an existing tree.

"Scion wood" refers to the previous year's new-growth twig, which we collect from the tree we want to propagate while the tree is dormant (in February or early March here in Vermont). The bottom or anchoring part of a tree is called the rootstock. Rootstocks are propagated by stooling. This means cutting a tree and piling up sawdust or compost around the base so that the new shoots develop roots. These shoots are then cut away to survive on their own, now called a rootstock. It's possible to use rootstock grown from seed, but these will vary genetically. Because rootstocks convey cold hardiness, height, disease resistance, and other characteristics into the tree, stooling is a good way to know what you're getting.

The general categories for sizing rootstock are dwarf, semidwarf, and standard. Within and between each of those categories are many different specific rootstocks. In general, the more dwarfing the rootstock, the more precocious the tree, meaning it will bear fruit earlier but not live as long. Rootstocks for taller trees bear later but have more longevity. We use different rootstocks for different applications. For example, our favorites for apples are Bud 9 for dwarfing (8–10′ / 2.4–3 m) and high-density plantings in our hoophouses; Malling 7 (M7) for semidwarfing and creating a manageable-size tree (12–15′ / 3.7–4.6 m) for homeowners through our nursery sales; and Antanovka for a cold-hardy standard-sized (20′ plus / 6.1 m) tree in our Knoll Orchard. In general, the more dwarfing a rootstock, the more babying it will need in terms of staking (or trellising) and irrigation due to its shallow, weaker roots. The cold-hardy rootstock we use for grafting plums is American plum, and for tart cherries we use *Prunus mahaleb*. Both will grow about fifteen feet (4.6 m) tall.

The main idea behind grafting is to join the actively growing cambium of the scion with that of the rootstock. We use bench grafting to put together trees to grow out for our nursery production. Our favorite bench grafting technique is called "whip and tongue," whereby we make diagonal cuts on rootstock and scion wood of similar diameters and join them together with the cambium layers of each lined up. The "tongues" are two notches that line up to give the graft more structure and support. The graft (joined part of rootstock and scion wood) is held in place with rubber bands and kept moist by wrapping with Parafilm. Many variations to this grafting method exist. We keep the trees cool in buckets of damp sand in the basement until the soil conditions are good for planting.

During a short window of time in the spring, as trees are almost coming out of dormancy, we can topwork them. This means we can add or change varieties of an existing tree. For example, scion wood can be added to convert branches to another variety. It can also mean cutting off most of the tree and changing it over completely to another variety. Different types of cleft grafting are our choice when topworking young rootstock "trees" that were planted the previous year. In this technique, we split the rootstock, cut the scion to a sharp wedge, and

FIGURE 7.4. A topworked apple tree shows the grafted scion wood. One new branch will be chosen to keep.

match up one side of it with the cambium on the outside of the trunk or branch. Plenty of soft grafting wax is used to seal up any cut surfaces. This can also be done on larger full-grown trees. Occasionally a grower will convert entire blocks of orchards with this top-grafting technique. We have also tried bud grafting. We take buds from the variety we want and insert them into a slit in the cambium of the rootstock. However, the best time for bud grafting is in the late summer, which is a tough time for us. We're really busy harvesting fruit then!

Apple Trees

We keep wondering why we didn't plant apple trees when we first moved to Vermont. John's master's degree research dealt with apples, and while still in Michigan, he dreamed of having his own orchard. It was almost fifteen years after we bought the farm that we put in our first apple trees, twenty-four of them (with several different apple varieties for cross-pollination and diversity) for the Front Lawn Orchard. We've been making up for that delay by planting a lot more. And pear trees, plum trees, and cherry trees too.

The current apple-growing paradigm is that you can't grow them without pesticides. Even organic apple growers often use multiple sprays of copper and sulfur to deal with diseases, and broad-spectrum insecticides, like Entrust or neem oil, for their war on insects. The long-term ecological effects of these chemicals on beneficial soil and leaf fungi or beneficial insects are not understood. We think there are some, which is why we don't go that route.

Back in the 1980s when we were in graduate school, there was a big paradigm shift for apple growing called integrated pest management (IPM). In theory, this meant primarily using cultural controls such as disease-resistant varieties, pruning to open up the canopy, and biological controls like releasing predatory ladybugs for mite control for apple pest management. Scouting was the mainstay of IPM and used to count pest densities for determining economic thresholds that could then be used to determine when a chemical spray might be needed, rather than relying on a calendar-based chemical spray regimen. IPM could reduce the amount of spraying in a typical orchard, but the underlying mind-set was and still is, "You need to use pesticides in one form or another to grow fruit commercially."

We questioned this apple-pesticide dependency. The use of pesticides emphasizes getting the maximum crop from the trees and places a lot of expense, effort, and external health and environmental costs into reducing the risk of loss. We can understand the dilemma, because the typical apple orchard grows only one crop and receives low commodity prices for it, hence the reliance on pesticides. Consumers also expect blemish-free fruit.

In our farming system, we put much less effort (and money for sprays and labor) into our crop management and take the yields that we get. We focus on proactive strategies rather than reactive ones. For example, we put in many varieties that are resistant to diseases such as scab, rust, and fire blight, the most common apple diseases in our area. We nurture and prune them and have marveled at the bountiful apple crops the trees produce.

Our orchards are small enough for us to hand-thin the apples when the apples are smaller than golf balls at times when there is too large a crop set. When thinning, we choose damaged fruitlets to remove and hot-compost them. This helps improve the size and quality of the remaining apples and evens out the production over the years. Without thinning, apple trees tend to produce biennially rather than annually. Conventional orchards use an insecticide spray that also acts as a fruit thinner to do this job.

We do get a fair number of "ugly" apples, which may take the form of blemishes from disease and insects and even the proverbial worms

FIGURE 7.5. Our organic Liberty apples were sold at the farmers market.

in the apples, but that's okay with us. We sort the fanciest apples for fresh market, using the dinged-up, gnarly fruit for cider and cider products, and the worst get hot-composted, including the drops. We collect the apple drops on the ground in the fall to reduce insect pests. Pests such as apple maggot and codling moth can mature in the rotting fruit and overwinter there. They will show up bright-eyed and ready to roll into the next growing season unless the apples are removed and composted or fed to animals such as pigs.

Our approach is a low-input, low-cost production system with lower yields than those of our conventional apple grower friends. We asked ourselves why we would spend 50 percent more effort and cash to get only a 20 or 30 percent increase in yield. Our profit comes from the difference between the low cost of production and finding more high-value uses and higher-value markets thanks to being pesticide-free. We prefer this minimalist approach.

We have taken this thinking to another level and found a few abandoned orchards that we work in to harvest apples with zero inputs other than harvesting labor and a little restorative pruning. Many of these apples are spotless, many are scabby, and some trees are blank every second year. Again, that is okay because we have little to no investment.

We have even started to think of apple scab as a possible value-added disease. Apple scab is a fungal disease that causes patches of rough corky skin on an apple. It can also damage the leaves and cause leaf drop when the tree has a bad case. Yet it has been shown that stress on the fruit enhances the production of secondary plant compounds for defense. These compounds may give more nutritional value to the fruit—an interesting concept. In any event, if we have a poor apple season due to a late frost or other phenomenon such as a disease outbreak, we still have many other fruit species to pick up the slack. That is the beauty of diversity over monocropping.

TABLE 7.1. Our ten favorite apple varieties.

Apple Variety	Description
Ashmead's Kernel	A russeted orangish mid-season apple. Ashmead's Kernel is a multipurpose apple with a rich balanced flavor that tastes a bit like a pear. It stores well, and has some disease resistance. Late-season apple.
Chestnut Crab	Technically a crab apple, the Chestnut Crab produces golf ball–sized orangish "apples" that are tasty, juicy, and sweet, like a Honeycrisp with attitude. We eat them fresh or blend them with other apples to make cider. Disease resistant for scab, fire blight, and rusts.
Duchess, and Red Duchess	Duchess of Oldenburg is the best pie and sauce apple ever. Hardy to zone 3. A tart, juicy apple, it ripens late August, though its storage time is short. Red Duchess is a selection with redder fruit (less yellow striping).
Empire	The Empire is a cross between a McIntosh and Red Delicious, though it is much tastier than either of its parents. It is a mid- to late-season ripener and stores well. Some susceptibility to scab. Moderate resistance to fire blight and rusts.
Gold Rush	Gold Rush is one of our favorite apples in terms of taste and storage, but we've had some problems with the trees' longevity on M7 rootstock on the farm. We're currently trying it on different rootstocks. Gold Rush is harvested late in our zone, mid-October. We've stored these until April in our refrigerator. The skins dry out a bit, but the flesh is still juicy and sweet. Yellow with a slight red blush.
Haralson	A University of Minnesota variety that also grows well in Vermont. A tasty red apple that is disease resistant for scab, fire blight, and rust. Good fresh eating and processing. Ripens mid-season.
Honeycrisp	Big, tasty, sweet red and yellow apples that we've been growing in one of our greenhouses as well as outside. Mid-season apple. Customers often ask for this variety, but it doesn't store so well as other varieties.
Liberty	One of our overall favorite varieties grown on the farm. Disease resistant for scab, fire blight, and rusts. Very prolific and tasty crisp apple. Ripens mid to late September.
Northern Spy	Late-season pinkish red apple that is tart and crisp with a complex flavor. Good for cider blends and pie. Heirloom apple with some susceptibility to fire blight and rusts.
Redfree	Ripens late August to early September. Juicy sweet apple. Good disease resistance for scab, blight, and rust.

When the trees in our Front Lawn Orchard were young, we mowed around them or down the rows. Perhaps because we were raised in the suburbs, something inside us seemed to default to lawn over the savannah look. Now that the trees are full grown and their canopies have grown into one another to some extent, the orchard is difficult to mow. For most of the summer, it looks a bit abandoned because of the tall grass. Then in August, John (and Sarah, if John feels like sharing his favorite job) goes out with the trusty scythe to cut the grass. This makes it easier to move among the trees when picking as well as making it easier to find any drops.

Apple trees took thousands of years and thousands of miles to travel from the apple-forested hills of Kazakhstan, where they are believed to have originated, to the maple-forested hills of Vermont. Like the European honey bee, earthworm, and dandelions, they were brought to North America by European colonists and to Vermont in the eighteenth century. Nearly every farm and homestead had apple trees and often a whole apple orchard. Some of these were seedling trees, but many were grafted trees of known varieties. There were apples for fresh eating, for cooking, for drying, for storage, and to fill the cider barrels in the cellar. Apples, including hard apple cider, were key components of the nineteenth-century Vermonter's diet.

We think about this connection our Vermont predecessors had with their food, especially apples. They knew so much about different varieties, their flavors, storage capability, and uses, as well as the cold hardiness and disease resistance of their trees. Most of this knowledge, as well as the varieties themselves, have been lost.

Renewed interest in making hard cider at home, as well as new cidery entrepreneurs, has prompted a whole new crop of apple enthusiasts. Hard cider apple varieties are having a resurgence in nurseries, too, for the "grow-it-yourself crowd." Orchardists are planting blocks of heirloom trees to meet the needs of the growing cider market, and hard cider workshops and talks are filled to capacity. Orchardists and other Vermont nurseries like ours are grafting and selling old and new apple tree varieties, including those for hard cider. We have visions of our Knoll Orchard eventually being like an old-fashioned cider orchard.

Pear Trees

We wonder why there aren't more pear trees growing in Vermont. Pears were also brought over by the colonists, and like apples, they're native to Europe and Asia. They also tend to be less susceptible to pests and diseases than apples. They, too, make a great hard cider, called perry. We especially like a blend of pears and apples in our homemade hard ciders.

So why aren't there more around? Well, they do take a bit longer to get established: "Plant pears for your heirs," as they say. That might be one of their drawbacks. They take a long time to bear fruit too. Another reason is that most pears need to be harvested while the fruit is still firm and not totally ripe. They're then ripened indoors, which, given the space requirements, may have been a big drawback for our Vermont ancestors trying to grow them.

We've put in a couple dozen trees within the past eight years. The first we planted have now started to bear a few pears, usually well out of reach. Pears get tall quickly, and they tend not to bear until they're a pretty good size. When mature, they'll be about twenty to thirty feet (9.1 m) in height. We've selected cold-hardy varieties for our Pear Corner Orchard including Early Gold, Flemish Beauty, Golden Spice, Parker, Stacey, Summercrisp, and Waterville. Most pear trees are not self-fruitful but require cross-pollination from a different variety that blooms at a similar time. The Flemish Beauty variety is self-fruitful.

Stone Fruit: Cherry, Plum, Peach, and Apricot Trees

We were spoiled in Michigan, where homegrown and You-Pick operations for cherries, plums, peaches, and apples were easy to find. When we moved to Vermont, we wanted delicious homegrown fruit in the summer as we'd had in Michigan. And we wanted it to be organic—which it wasn't in Michigan. That's why the first fruits we planted on the farm were two tart cherry trees and two plum trees. We're a solid zone 4a (USDA Plant Hardiness Zone) in our part of Vermont, and though with climate change we've experienced warmer winters and summers, we can still get a few negative 30°F (−34°C) nights (or lower) in the middle of winter. It's that lowest temperature that dictates what we can grow. When we first started growing, peaches and apricots were pretty much out of the question, although nurseries have since introduced new hardy varieties and our winters seem to be getting milder. So far, though, no peaches have survived on our farm more than two years outside the hoophouse.

Stone fruits also like full sun and well-drained soils. The main challenge with stone fruits here in Vermont is that they tend to bloom in early to mid-May. Since our frost-free date isn't until May 31, the flowers are susceptible to spring frost damage. Apples and pears tend to bloom later in May and are usually not impacted by an occasional late-May frost. We don't get tart cherries or plums every year, because of the frosts, but when we do, they are such a treat.

Another issue with stone fruit in Vermont is they tend to survive for only about twenty or twenty-five years, as they're susceptible to viruses that eventually do them in. As if on cue, our original trees died within that timeframe. We had a few bumper crops, a few decent crops, and a few years with nothing at all because of May frosts. We've since planted many more.

Southwest disease can be common on stone fruit trees. When the trunk gets exposed to bright, sunny winter days, the trunk warms. If such days are followed by freezing cold nights, the warming and freezing can cause the trunks to split. This in turn weakens the tree and opens up places for disease-causing organisms to enter. The first two cherry and two plum trees we planted were shaded in the winter because of a row of white spruce on their south side and the low angle of the winter sun. They never showed signs of the southwest disease. Several cherry trees we planted about eight years ago, however, are now in decline because of that very problem.

We learned the hard way about the importance of trunk wraps in the winter for these exposed trees. White plastic trunk wraps help ward off the winter heating problem, and they also prevent vole damage. Just remember to remove them in the late spring and early summer; as with apples, if the wraps are not removed, they will keep the trunk moist, making it susceptible to disease and insect damage. While stone fruits are most susceptible to the southwest disease, it killed one of our six-year-old apple trees too. Did we mention that fruit trees can break your heart?

Tart cherry trees (the only kind of cherry trees cold hardy enough to grow here currently) are self-fruitful, which means only one tree is needed for pollination. "Self-pollinating" and "self-fertile" are other common terms that mean the same thing. Most sweet cherries need a pollinating partner of a different variety, although there are a few self-fruitful sweet cherry varieties available. European plums, peaches, and most apricots are also self-fruitful. With that said, they tend to have better pollination when different varieties of trees of the same species are nearby. European plums come in an assortment of colors and flavors and tend to be oval in shape. They're a bit more susceptible to black knot than the American versions.

TABLE 7.2. Stone fruit grown at The Farm Between.

Tree Species	Variety	Description
Apricot, Manchurian (*Prunus mandshurica*)	Brookcot Scout	We are growing these in an unheated hoophouse, where they have overwintered well but have not flowered yet. They are rated zone 3 hardiness and are self-fruitful. A few Manchurian apricot seedling bushes along our fencerow have been growing for five years. A few fruits have formed but disappeared before they ripened in July (thanks to chipmunks and squirrels).
Peach (*Prunus persica*)	Reliance	Reliance is rated to zone 4, but the trees we've planted in different years have eventually winter-killed. Currently, we have a couple growing in our unheated hoophouses. They are growing well, but unfortunately, no peaches yet.
Plum, European (*Prunus domestica*)	Mount Royal Greengage	European plum varieties are self-fruitful and come in different colors and shapes with a hardiness zone range of around 4–8. We've been growing these for only a few years—but so far, so good. The plums are delicious.
Plum, Japanese-American Hybrids (*Prunus salicina* × hybrids)	Alderman Pipestone Superior Toka Waneta	These trees all require another variety of Japanese-American or American plum for cross-pollination. Zone 4–8 hardiness with good results in our area. Bloom times vary, so if you plant only two trees, make sure their blooming times overlap. We match Alderman with Superior or Waneta, and Toka with Pipestone. All are delicious. Toka tastes sweet like bubblegum. Toka and Waneta have had smaller fruit but are more prolific for us than others. Ripen late August to early September.
Plum, Native American (*Prunus americana*)	Seedling Trees	Native shrubby trees have delicately sweet-smelling blossoms in the spring that attract dozens of different native pollinators. Heavy producers of small plums that ripen in August. Great for jellies and processing. Variable for fresh eating, as the skins can be tannic and sour, but inner flesh is tasty. Great tree for cross-pollinating Japanese-American hybrid plums.
Tart Cherry, also called pie or sour cherry (*Prunus bali*, *Prunus cerasus*)	Evans Bali Mesabi Meteor Montmorency North Star	General hardiness for these self-fruitful tart cherries is zone 4–7 except for Evans Bali (zone 3–7). Growing well on our farm, although trees have been susceptible to southwest damage and different viruses that have limited tree longevity. Fruit ripens in July. We like to eat the cherries right off the tree, and process them for juice, jams, and pies.

Many of the stone fruit trees we grow are known varieties grafted onto hardy rootstock. Japanese-American hybrid plums require a pollinating partner of a different variety for cross-pollination. Plums from Japanese-American hybrids are the reddish or maroon round plums with yellow to orange flesh that are generally found in the supermarket. Many different Japanese-American varieties are available that are hardy to zone 4.

We also grow the American plum, a native that can be a good pollinating partner for Japanese-American hybrids. If you want to grow only American plums to make jelly, or wine or beer, then you will need two different seedlings for cross-pollination. As with apples from seedling trees, plums from seedling trees can also vary. One tree might produce fruit that is sweet and juicy, while another might produce tannic or mealy fruit.

We've also planted a few apricot and peach trees around the farm, but so far haven't had too much luck keeping the trees alive, as the river valley where we live tends to get very cold in winter and the winter winds are dry, which may be an issue for these types of trees. However, they're overwintering well in the protected hoophouses.

FIGURE 7.6. American plums are favored for making delicious jelly.

Mulberry

Years ago we planted two Russian mulberry trees, hoping to raise silkworms to use in Nancy's fiber projects. Silkworm larvae feed on mulberry leaves, and these trees were brought over in colonial times for silkworm production. Mulberry trees also produce delicious fruit that look a bit like blackberries. The tree is now considered naturalized in the Northeast. It is grown from seed, unlike grafted fruit trees.

One of our first mulberry trees suffered root rot, *Armillaria mellea*, a fungus that attacks the roots and causes leaves to drop prematurely. It can also cause the death of branches or even the entire tree; our tree died after a few years. Our second mulberry tree died after it split in half during a windy winter day. It may also have had root rot. We did have an amazing crop of mulberries the summer before it died. The berries ripened over a period of about a month. We'd stop on our daily walks and eat our fill from the lower branches while the birds sat on top eating theirs. Quite a treat for all of us.

We recently planted a couple more trees, far from the original two because the fungus can spread via mycelia in the soil. As we already mentioned, fruit trees can break your heart and lead you to question whether "'Tis better to have loved and lost / Than never to have loved at all." We'll take the first option every time.

Nut Trees: Hazelberts, Black Walnuts, and Chestnuts

At Christmastime when we both were kids, our families would buy a bag of assorted nuts with the shells on; almonds, pecans, Brazil nuts, walnuts, and filberts. Our moms would put them into special wooden nut bowls that also held the nutcracker and little metal picks to pick out the bits of nut that stayed in the shell. We had to work for our treats in those days. Another special treat at Christmas was roasted European chestnuts. Now, those were good. Of course we want to grow nuts!

Unfortunately, northern Vermont's climate doesn't allow us to grow all those favorite Christmas nuts, but we can grow a few. Hazelberts are a close relative of filberts and make a great nut tree for the home or farm. A cross between the American hazelnut and European filbert, hazelberts have both the cold hardiness of the American hazelnut and disease resistance to the Eastern filbert blight. Hazelberts generally bear larger nuts than those of the American hazelnut.

FIGURE 7.7. Hazelbert nuts ripen in their husks and are ready when the husks turn brown.

These shrubby trees grow to about twelve feet (3.7 m) or so. Because they sucker, they grow into a great hedge that turns a coppery color in the fall. Suckering is an asexual vegetative reproduction process, common to lots of plants, whereby they send out new stems from their root systems. Raspberries are a good example of a suckering plant, as are hazelberts. One such beautiful hedge hides the back sides of our hoophouses from our neighbors. For the trees to produce nuts, you need at least two seedlings for cross-pollination. This means you can't just dig up a bunch of suckers from the same tree and move them to a new place and expect to get pollination. The nuts are ready in early fall when the husk turns brown. Pick them before the squirrels and chipmunks take and hide them all.

Back around 2010, we planted about fifty black walnut trees in the pollinator sanctuary as part of our north-south hedgerow windbreak wildlife corridor. We lost a few that first year, but many have survived. Today, they're fifteen feet tall (4.6 m) and beautiful, although they'll probably need another thirty years to be harvestable for their sought-after veneer wood. Maybe our kids or grandkids will profit from that, but we should get nuts much sooner. Black walnuts are good to eat and have a strong flavor. Their husks can make a chocolate-brown dye for both wool fibers and your hands if you don't wear gloves.

Some people might wonder why we've planted these black walnut trees if we're not going to derive the benefit of the harvest. It's true that we're not going to see them fully mature, but like grandkids, we can still enjoy them while they're young.

We've also tried planting several Carpathian walnut trees, a more cold-hardy strain of the standard English walnut tree, but whose nuts taste similar. Carpathian walnuts have been rated to survive to −30°F (−35°C), so their chance of success is iffy in our region. Of the four we planted, only one lasted for several years, but died back to the ground in the winter of 2018. It has resprouted from its roots and hope springs eternal, so we haven't given up.

Chestnuts used to be common in southern Vermont until the chestnut blight pretty much wiped them out. We have planted American chestnut, Chinese chestnut, and hybrids of the two to see what they can do here. The nuts from the American chestnuts are smaller and supposedly sweeter than the Chinese or European chestnuts. The American-Chinese hybrids are bred to have the chestnut blight resistance of the Chinese trees with the sweeter nuts of the American. We've tried nuts from hybrid trees at our local farmers market, and they are tasty. The Chinese chestnuts we planted died from winter damage, but a few of the hybrids are hanging in there. We'll keep trying because we would love to have homegrown chestnuts in the future! They also make a nutritious gluten-free flour.

FIGURE 8.1. Black currants are just about ripe for picking.

CHAPTER 8

uncommon berries

Make no mistake about it, we enjoy growing, eating, and marketing common fruits such as raspberries, blueberries, and strawberries. We started growing highbush blueberries in the early years on the farm and added more later. Blueberries can be a bit finicky because they require rich, well-drained, acidic soils. Our blueberry site gets a lot of frigid winds in the winter, too, which has also been an issue for us in terms of occasional winter damage and loss of flower buds. Raspberries, as mentioned earlier in the repurposing unheated hoophouses section in chapter 5, have been a great crop and a financial success.

Our true berry passion, however, has been embracing the production, processing, and marketing of lesser-known cold-hardy fruits such as currants, gooseberries, elderberries, aronia, and more. This is the niche we are best known for, and one that is sparking the interest of the growing artisanal beverage industry in Vermont. In this chapter, we discuss our favorites.

Black Currants

Black currants originated in Europe and Asia and were brought to North America by the early colonists. They are cold hardy, adaptable, and even grow well in partial shade. They like cool, moist soils and lots of mulch. The European black currant has become one of our favorite fruits for eating and growing, although that wasn't always the case. When we harvested our first black currant crop back in 2004, our faces puckered whenever we popped one of the black beauties into our mouths. But over the years, we've learned to wait until they're really ripe and juicy before we eat them. They taste a lot better that way. Our taste buds have adjusted as well. Now we recognize their complex tart, savory, and tannic flavor as a pleasurable treat, and we look forward to the end of July, when we can eat the plump berries right off the bush.

Black currants are filled with eighteen times more vitamin C and twice the amount of antioxidants than blueberries, which are often touted as a superfruit. Compared with blueberries, black currants also have nine times the calcium, and four to five times the iron, magnesium, potassium, phosphorus, and vitamin A. Now, that's a super superfruit!

Black currants also make delicious syrups, sauces, jams, sorbets, wines, and more. A local winery buys the majority of our crop, but we occasionally sell them to breweries and cideries too. They are definitely becoming more mainstream as time goes on. That they are relatively easy to grow and are pretty much "deer-proof" and somewhat "bird-proof" makes them an almost ideal fruit. No wonder the early colonists brought them over.

But here's the flip side: Until recently, black currants (and other species in the genus *Ribes*, such as gooseberries and red currants) were banned in most of the United States because of concerns about a fungal disease called white-pine blister rust. Currants and gooseberries are the alternate host for the blister rust fungus, which means these *Ribes* species are needed by the fungus to complete its life cycle. White-pine blister rust also originated in Europe and was inadvertently brought to the United States in the late 1800s. Because American white pines (and other related pines) do not have disease resistance to the rust (as many European pines do), they tend to die within a few years of contracting the disease. Thus, a ban was enacted in the early twentieth century to protect the pine industry. The black currant was believed to be the more serious culprit, so in states where there is a white pine industry, they might still be banned.

Many states, such as Vermont and New York, have lifted the ban because studies indicated that the woods are already populated with "wild" or escaped *Ribes*, thus the threat from homeowners or commercial growers is insignificant in comparison. Also, as blister rust–resistant black currant varieties, such as Titania and Consort, have become more widely available, much of the concern has gone away. There is, however, recent documentation that the rust organism is evolving to break down this varietal resistance farther south of us, so we are keeping an eye on things.

Back in 2002 we obtained a SARE grant to experiment with black currants. We conducted a variety trial and looked at overall yield from three different black currant varieties: Titania, Ben Lomond, and Ben Sarek. Titania, by far, had the highest yields and best plant vigor. Both Ben Lomond and Ben Sarek were susceptible to white-pine blister rust. Our yields for Titania ranged from an average of 2.5 to 4.7 pounds (1.1–2.1 kg) per plant in years three and four, respectively. Since those days, we've incorporated other varieties, including Consort and Minaj Smyriou. Both are vigorous, good-tasting black currants with cold hardiness and rust and mildew resistance similar to those of Titania.

If you have only a few black currant bushes, they're easy to pick. Just make sure they're nice and ripe so you can pick handfuls instead of a couple at a time. A few bushes can be fun, but we have several hundred, which requires a lot of picking labor. It seems as though every year it gets a little harder to spend the long hours in late July and August needed to bring in the black currant crop. A few hours into picking, and we're already feeling our backs, our feet, and on those hot humid days, our eyes stinging from dripping sweat. One year, Nancy picked about three hundred pounds by herself, one berry at a time, for eight-plus-hour shifts, but those glory days are past now. Lately, after more than four hours of picking, she cramps up at night and can't sleep. But still she puts in those long days. Luckily our crew likes to pick. We also recruit local labor that we pay by the pound. They help pick up the slack for us, if needed.

To harvest currants, we use plastic one-gallon milk jugs with their tops cut off but their handles intact as quart container holders. Tied around the waist with baling twine, they allow us to pick with both hands. It works great, but we can't always hit our target of picking nine to ten pounds (4–4.5 kg) per hour.

Some years we've also had help harvesting currants from kids attending local camps and other summer programs. They got to pet the horses, a free tour, maybe a snow cone, but they gave back as well. We'd tell them they can eat as many black currants as they want, which is a standing joke since kids don't usually like black currants much. We kept those hungry hordes away from the blueberries and raspberries, though! Once we had a farmer training group from Quebec spend the night on the farm and help pick currants the following morning. They really cleaned up the bushes. They ate a lot, too, but it made us happy to have others around who shared our love of fresh currants.

There's an art to picking black currants and other berries. The trick is to move and manipulate the branches rather than moving around a lot yourself. It requires focusing both your energy and your intent. A couple of the most physically fit, athletic people to ever help us pick turned out to be inefficient; they moved their bodies too much, twisting, bending, weaving, turning, as if it were a sport requiring maximum physical effort. We tried to tell them to stand still, to move the branches full of hanging berries in order to quickly grab them by the handful. But they were athletes, and they approached the task as athletes: "Move your body, then move your body faster." Not an efficient way to pick berries.

So-called spot pickers, who spend a lot of time moving and looking for a better spot when actually it's right in front of them, are also slow. Systematically picking up branch after branch and collecting all the berries in that one place is really the quickest way to do it. And we harvest most of the berries that way. Our current currant crew, Alisha and Sarah, get it, and are quite good at it.

At some point in the long picking session, we reflect on the pickers of the world. They might be tea leaf pluckers, coffee bean pickers, vegetable pickers, or berry pickers like us. Many of the pickers of the world are women, almost all are underpaid, and they endure very difficult, often unsafe working conditions. If we put in eight hours of picking in a day, we think it's a big deal. For them, ten or twelve or more hours are expected, every day. They say you have to walk a mile in someone else's shoes (or maybe they don't even have shoes) to have an understanding. But even then, you can't really understand someone else's story. Thinking about the other pickers of the world while we pick doesn't really give us understanding of their lives, but we do get a small sense of solidarity. And it gives us an even greater appreciation and gratitude for our farm and the relative ease of our farmwork and lifestyle.

Clove Currants

The clove currant is one of our favorite spring and late-fall berry bushes. This black currant is native to the Midwest (the Northern Great Plains of the United States and north to the Canadian Prairies), and distinct from European black currant varieties in both taste and appearance. In May, its small, yellow, trumpet-shaped flowers smell sweetly of

FIGURE 8.2. Clove currants make great edible landscaping plants.

cloves. We love it, as do the bumble bees. It's a pretty bush, too, about three to four feet (0.9–1.2 m) tall, with light green three-lobed leaves that turn a beautiful pinkish orange that persists into the late fall.

In Vermont, the large shiny clove currants are ready to pick in late July and early August, after most of the other *Ribes* have been harvested. They make a nice treat at that time of year, their taste and texture more like Concord grapes or gooseberries than currants. Their skins are a bit thick, but they're delicious fresh eating and in jams and pies as well.

The two biggest challenges with this type of currant are that their branches get so heavy with fruit that they tend to flop over, and the fact that they don't all ripen at the same time, which makes them a bit harder to pick. We've pruned out the lower branches to keep those off the ground. Regular pruning helps invigorate the plant and gives it a desirable landscaping and a shape that makes it easier to pick. We've also used some wooden stakes to keep it upright, which have helped in picking. Although we haven't tried it, these bushes might be good candidates for trellising when used for a serious commercial venture. We've been using them and recommending them to homeowners more for their landscaping appeal than high production but wouldn't rule them out as a cash crop.

Red Currants

Red currants also include white and pink cultivars. All are hardy and easy to grow. When grown in Vermont they prefer partial to full sun; in hotter climates they might desire even more shade. Their beautiful clusters of red, pink, or white berries hang on stem structures called strigs. Cedar waxwings, gray catbirds, and other birds tend to eat them when they're starting to ripen, so we net ours early. We throw the nets over hoops and use lots of metal staples to keep the net anchored to the ground. The birds are pretty crafty and will find any loose gaps, so we throw old bricks on the edges for good measure. If you don't tighten and secure the net at the bottom, or if the net has any holes, the birds will get in—then you have to get them out. The best way to free them is to open one end of the net and chase them out by starting at the other end. Then it's important to get extra crazy securing the bottom or sewing up any holes. Once they've had a taste, they will try anything and everything to get back in. We've tried holding and scolding them before letting them go, but that trauma seems to be quickly overcome by the promise of another mouthful of berries.

FIGURE 8.3. Red currants, Rovada variety, are ready to pick.

Berries ripen from the top of the strig down. The top berries are also the biggest. Many people have told us they get only a few small berries from their bushes even from varieties that produce big berries. Most likely the birds (cedar waxwings specifically) are out there at the crack of dawn, eating all the big, partially ripe berries. Cedar waxwings tend to send scouts to find fruit and then move around in groups. When they find good berries, they'll bring all their friends. On a few unnetted bushes, they seem to leave the smaller red currants even when they're red and ripe. Whether that's because they've moved on to other berries on the farm or because by then they have turned their focus to nesting and feeding insects to their young, we don't really know. They tend not to hang around the berries so much in late July and August, during their nesting time.

With red, pink, and white currants, it's best to pick the whole strig if you want to keep them in the refrigerator or to sell later. They'll hold up for a week or more in the cooler or fridge that way. Once they've been pulled from their individual stems, they tend to shrivel and don't look so appetizing, although they still taste fine. If you're going to freeze them right away, it's fine to pull them in handfuls off the strigs as we do with black currants. Gooseberries and black currants hold up for a week or two in the cooler, but they don't require staying on the stems as red currants do.

Currant season is the time we tend to meet a lot of Europeans who live in Vermont; Germans, Scandinavians, Eastern Europeans, and Russians. They're the ones who stop at the farm or farmers market booth to buy the beautiful red and black currants. The typical American tourists might try a sample, but their puckered faces give them away. They are more likely to opt for the sweetened syrup on a snow cone. Red currants, which also include white (albino cultivar) and pink varieties, are great for jelly, but we like them for fresh eating too. Pink champagne currants, in our opinion, are the tastiest of the red currants for fresh eating.

Nancy grew up on her mom's pretty raspberry-red currant jelly, which she remembers eating on her peanut butter and jelly sandwiches. Her parents sold fresh red currants at their fruit market, and their Swedish neighbors from three doors down had several bushes in their backyard. (They would let the neighborhood kids glean a few when they were done picking.) Nancy also grew up in a town that had a population where almost 50 percent of the people were of Swedish ancestry, which helps explain why red currants were available even though they were officially banned in New York State.

TABLE 8.1. Our favorite currant and gooseberry varieties (zone 3–7).

Variety	Notes
Black Currant—Minaj Smyriou (*Ribes nigra*)	Vigorous, cold hardy, productive with a bit milder taste than Titania. Resistant to white-pine blister rust. Similar to Titania in terms of picking, producing, and propagation.
Black Currant—Titania (*Ribes nigra*)	Most of the currants we grow and sell are Titania. They're easy to propagate, grow quickly, are good producers, and are resistant to white-pine blister rust. Delicious, easy to pick, freezes well. De-stem before freezing when using for jam.
Crandall or Clove Currant (*Ribes odoratum*)	Sweet clove-smelling yellow flowers in May. Plants grow well once established but are more difficult to propagate and are slower to establish than black currants. Big berries that do not all ripen at the same time. Very different flavor than *Ribes nigra*, reminiscent of Concord grapes. Pretty pink leaves in the fall that last into November.
Gooseberry—Black Velvet (*Ribes uvacrispa*)	Large dark purple berries with good taste. For fresh eating, make sure they are very dark. Makes a superb jam with no added pectin needed.
Gooseberry—Hinnomaki Red (*Ribes uvacrispa*)	This large red gooseberry originated in Finland. Let these get dark red or maroon before harvesting for fresh eating. Also makes a great jam.
Gooseberry—Poorman/Pixwell Type (*Ribes uvacrispa*)	These berries tend to be smaller and sweeter (if you let them ripen to a reddish color) than the others. The most productive gooseberry on our farm. For jams, pick when slightly underripe for their higher pectin content.
Pink Champagne (*Ribes rubrum*)	This is a red currant cultivar dating from colonial times. Medium-sized berries. Let them get very pink for better taste. Sweeter than red currants. These ripen after Jonkheer and before Rovada.
Red Currant—Jonkheer Van Tets (*Ribes rubrum*)	Red jewels that ripen early to mid-July. Medium- to large-sized berries. Let hang under the nets until all the berries on the strig are red. This will ensure they are juicy and tasty (although still tart).
Red Currant—Rovada (*Ribes rubrum*)	Ripens late July to early August. As with Jonkheer, let these berries hang under the nets to get nice and ripe. Very large tasty, tart berries. Plant with Jonkheer for a longer red currant season.

Gooseberries

Gooseberries tend to be a little bigger and sweeter than currants. There are so many interesting varieties now available to growers, each with a slightly different flavor. We love growing them for the wide range of taste sensations. Gooseberry bushes tend to be smaller than currant bushes, around three feet tall as opposed to three and a half to four feet for black and red currants. Gooseberry bushes appear green and lush in the spring, but by late July and August, they tend to look a little ragged, especially when conditions are hot and dry. Like currants, gooseberries in Vermont's cooler summers do well in full sun but would prefer partial shade in hotter climates. They also originated in Europe and grow well in these northern climes.

Picking gooseberries is a little harder than picking currants because they require you to bend over, squat, or else get down on your knees. We've seen some systems in Europe where growers trellis the bushes in a two-dimensional espalier fashion. This might be something to try when we have spare time! But the thorns are the biggest challenge. Picking gooseberries is a great way to practice mindfulness; their long, hard, sharp thorns can easily draw blood when you let your mind, and your fingers, wander. The trick to picking gooseberries is again to manipulate the branch to make the berries hang down and more easily accessible, rather than reaching your hand into the bush, which is a good way to get poked. You can even wear a glove on your branch-holding hand for extra protection.

Gooseberries make a phenomenal jam with big hunks of tart berries floating in the sweet-tart crimson jelly. As with the other *Ribes*, no pectin is needed, only fruit and organic sugar or honey. You do need to de-stem them, but taking off the remnants of their flowers from their tips is not necessary. The Irish call that "tipping and tailing." We have found that unlike the stem, the flower remnants get soft when cooked. We'll take it as extra fiber and less labor. We've made jam from all our varieties, but Black Velvet is our favorite. As the name suggests, you need to let these berries get very dark before they're ready to harvest. Gooseberries also go great with red currants to make a gooseberry–red currant syrup.

During the hot dry summer of 2018, many of our currant and gooseberry plantings dropped their fruit early. The bushes located in partial shade tended to hold on to their berries, while those in full sun dropped fruit during the July heat wave, when temperatures were in the nineties for over a week. That was hot for Vermont. We've since heard that a gooseberry grower in the Champlain Valley south of us had a total crop failure. Gooseberries really are a cool-climate fruit. The bushes themselves looked pretty ragged by the end of the summer. Most of the leaves had dropped or turned yellow. If you're concerned about aesthetics in your garden, this is something to remember before planting gooseberries. They look great in the spring but beat up in August and September. Climate change may require us to further adapt, including using shade cloth or irrigation to keep growing these delicious fruits successfully.

Elderberries and Aronia

We often think of elderberries and aronia together because they prefer similar growing conditions—that is, they don't mind an occasional flood and can do well in heavier soils, such as those in our flood-prone field 6 and other areas on the farm. They also both require full sun.

Elderberry and aronia are becoming popular with people who want to take charge of their own health care by using medicinal plants that fall outside of the Big Pharma pill-popping system. We, and our employees, eat them up as part of our "small farma health care plan." They're both high in antioxidants, vitamins A and C, minerals, and a variety of other plant compounds that are thought to exhibit anti-inflammatory, anticancer, and antiviral properties, to name a few. Modern scientific studies are finally catching up to what Hippocrates recommended more than two thousand years ago: "Let food be thy medicine and medicine be thy food." Elderberries, for example, have been clinically proved to be antiviral and reduce the time that people suffer from colds or the flu.

Historically, Native Americans used aronia for food, medicine, meat preservatives, and as dye plants. In the early 1900s, aronia was introduced into Russia and other European countries, where it has become an important juice berry, some of which is now imported to the United States. Likewise, elderberry and elderflower have been used for centuries for their medicinal qualities by Native Americans, Europeans, and European colonists. While elderflower cordials

FIGURE 8.4. Aronia berries are a native fruit with high concentrations of antioxidants.

and elderberry syrups fell out of favor with Americans by the mid-1900s, elderflower and elderberry continued to be an important crop in Europe and still are today. In Vermont, aronia and elderberry are garnering interest from the state's growing herbal medicine and beverage businesses.

Aronia berries and elderberries both freeze well, so they can be processed and used later. Once the fruit is frozen, you can shake elderberries off their clusters to more easily remove the stems, or when fresh, use a sieve to separate berries from the stems. Aronia can be eaten raw, although the berries are quite astringent. Both berries make excellent jams, jellies, wines, and syrups. Elderberries with their small berries are ideal for processing by steam-juicing.

More than a superfood, aronia and elderberry are also great for landscaping and wildlife. Aronia and North American elderberry cultivars are easy to grow, disease resistant, good berry producers, and cold hardy enough for Vermont winters. Although very different in size and shape, both make attractive landscaping plants. The waxy green leaves of the aronia bush turn reddish in the fall, making them an attractive native ornamental for around the home. This bush grows to about six to eight feet (1.8–2.4 m) and spreads slowly by suckering from the roots. Only one variety is needed for pollination. Common cultivars include Viking, Nero, and McKenzie. Native seedlings have also produced well for us with good berry size.

The American elderberry shrub grows to about twelve feet (3.7 m) and, once established, sends out suckers that allow it to spread and become bushy. A European subspecies (or species, depending on which taxonomist you talk to) of elderberry, *Sambucus nigra nigra*, is also being cultivated in the United States, although they haven't thrived on our farm.

Elderflowers bloom at the end of June through August depending on the variety. Their large white-flowered clusters give off a heady, sweet scent. They are mostly wind pollinated, and two different varieties are needed for cross-pollination. Insects will also visit the pretty white flowers for a sweet treat. We occasionally see honey bees visiting and wonder if they are self-medicating.

Elderberry clusters are called panicles. The berries generally ripen in August and September. Rather than picking individual berries, growers snip the panicles off with fingers or pruners. Most of the large stems should be removed before processing the berries into a juice or syrup. Our current favorite variety is called Marge, and is a cross between the American elderberry and the European elderberry. Unlike her European parent, which often dies back on the farm in

FIGURE 8.5. Alisha picks elderberries. The Bob Gordon variety is shown in the foreground.

winter, Marge does very well and is also easy to propagate. Our experience with Marge supports what the research shows: Marge is a vigorous, high-yielding, large-berried elderberry.

We are conducting an ongoing trial of six different elderberry varieties (Ranch, Wyldewood, Marge, Adams, Bob Gordon, and Nova). While Marge has been the top producer, both Bob Gordon and Ranch have also done well. Bob Gordon was bred to have a high brix (sugar) content. Ranch was selected for its more compact size, which makes it good for home growing and picking. It is more drought tolerant than other cultivars. Wyldewood does not seem to be a particularly hardy variety on our Vermont farm. It hasn't overwintered well here, and the stand is very spotty. We have grown all the more common varieties (Adams, Johns, York, and Nova) here on the farm, to varying degrees of success.

TABLE 8.2. Elderberry varieties grown on the farm.

Variety	Notes
Adams	Early-producing variety that has not been highly productive for us. Though Adams is often touted as being a good cross-pollinator for Johns, we now recommend varieties such as Bob Gordon or Ranch as a better choice.
Bob Gordon	Good producer. Berries have one of the highest brix contents. Sweeter large berries that ripen late August, early enough to avoid some spotted wing drosophila (SWD) pressure.
Johns	One of the traditional cultivars. Early-season berries. Often matched with Adams for pollination, but Johns also shares bloom time with Ranch, Marge, or Bob Gordon.
Marge	Our favorite variety thus far. This cross between the European and North American species is fast growing and vigorous, producing large juicy berries that ripen in the end of August, a week or two after Ranch.
Nova	Blooms later in the season into August with September berries. If we have early September frosts and cold nights that can reduce SWD populations, we can get good late-season harvests.
Ranch	One of the first varieties to ripen in mid-August before SWD pressure. Medium-sized berries on shorter shrubs that are good for backyard growers. Overall, good producer.
Wyldewood	Fast grower during the season, but didn't overwinter well the first few years on the farm. We don't encourage it yet for our area, as we want to see how it continues to fare.
York	Late-season berry. Similar to and good pollinator for Nova.

Unfortunately, in the last few years, elderberries have become a target for an invasive fruit fly called spotted wing drosophila (SWD), which we discuss more in chapter 10. This pest causes berries to drop before the whole panicle is ripe. Our method of dealing with SWD in elderberries is to pick the ripe portions of the panicle and pick every few days. This has worked fairly well, but our yields have declined compared with pre-SWD days. Because of this pest, we harvest a fair amount of the elderberry flowers and sell them wholesale. Before SWD, selling the flowers didn't make sense, as we could make more money from the berries. The economics changed quite a bit with SWD.

Because SWD builds up in late August, we generally prefer to harvest early-bearing varieties before the pest populations get too high. However, when we've had cool weather in September that has knocked down SWD populations, the later-ripening York and Nova (which mature in late September) have also done well. These later blooming and bearing varieties have also been used for their flowers. Both Adams and Johns varieties mature early. Adams is typically touted as a good cross-pollinator for Johns, which tends to be the better producer. Adams has not grown very well on our farm, especially as compared with Marge and Bob Gordon, so we are moving away from that variety.

A lot of cheaper nonorganic elderberries are coming out of large plantings in the Midwest. Although many conventional brewers and cider makers buy them, many small-scale users making medicinal products and local brewers remain committed to buying organic.

We have a precautionary note to add about elderberries. They should be heat treated or fermented before you, or others, consume them. The American subspecies has cyanide-inducing compounds that can lead to nausea, vomiting, and diarrhea if the fresh berries are consumed in sufficient quantities. These compounds are concentrated in the stems, so a few raw berries won't bother you, and in our opinion, you'd be hard-pressed to eat more because they don't taste that good straight off the bush.

Because of their ability to withstand occasional flooding, wild elderberry bushes are often found on the upper banks of streams and rivers, which makes them good plants for riparian zone restoration. This in turn can help protect water quality in our streams and rivers by reducing bank erosion. We've planted ours in marginal soils and the flood-prone areas on the farm, where they are thriving. They don't, however, tolerate consistently wet spots as much as aronia does.

Aronia berries don't seem to be a favorite food for insect pests (including SWD) or birds, except the first year we grew them. There were only a few berries per bush at that point, but right before we were about to harvest them, the cedar waxwings came through and picked our small bushes clean. In subsequent years, we haven't had a problem. We don't have an explanation for everything!

Honeyberries

We've been pleased with honeyberries (a.k.a. haskaps, or whatever you want to call them) especially as a homestead and personal-use planting. These plants originated in Siberia and northern Japan, where they are very popular. The honeyberry (along with saskatoons) are the first fruit plants to bloom on our farm. In early May, the pretty

uncommon berries

yellow honeyberry flowers burst open and attract loads of bumble bees. We love to walk through the rows, marveling at the dozens of bumble bee queens recently emerged from their overwintering sites. We can usually identify five or six different bumble bee species on the honeyberries. There are also solitary nesting bees of assorted sizes and colors that appreciate the honeyberry flowers. If the weather cooperates, May is a great time to be a flower.

Honeyberries are elongated, blue, and a bit tarter (and more flavorful) than blueberries. They make one of the best jams we've ever tasted, which was always a big hit at the farmers market. Their only drawback is they are a bit harder to pick than most other berries because they tend to hide in the foliage. In June, we keep a sharp eye on the berries to make sure we get the bird nets on before they start turning blue. These berries are a favorite of cedar waxwings, which migrate annually back to the area and seek them out, still hungry after eating all the bitter highbush cranberries that hung on all winter. We net early because once the birds get a taste of the drooping dark blue honeyberries, they will do anything to get them.

It's best to wait until the berries are dark purple all the way through before picking. A good way to tell if they're ready is when

FIGURE 8.6. Honeyberries (haskaps) ripen in June.

they start falling off or fall off when touched. Another good way to tell when they're ripe is if you start eating them and don't want to stop.

Honeyberries belong to the honeysuckle family, are non-native, and cold hardy as heck (to USDA zone 2). The orange-brown bark on their stems looks great in winter and early spring. The leaves are a smooth soft green that stay lush throughout spring and summer. These attractive bushes grow four to six feet tall (1.2–1.8 m) depending on the variety. Two different varieties are needed for cross-pollination, although a few varieties are so closely related that they don't cross-pollinate well (for example, the Indigo Series with Tundra and Borealis). It's important to find the ones that not only bloom at the same time but will also cross-pollinate each other. We currently grow Aurora, Borealis, Tundra, and Berry Blue (Czech 17) varieties. Borealis berries tend to be a little smaller and more bitter. Aurora and Berry Blue are big with a good flavor. Tundra tends to produce a lot more foliage, which makes them a bit harder to pick, but they are also a good size.

After we've picked our fill, we take off the nets and let the cedar waxwings, catbirds, robins, and other birds pick the remaining few as well as the drops, an excellent way to prevent future pest problems. After a couple days, they've done their job—there isn't a berry on the bushes or the ground. Now, that's a good cleanup crew, and they work for food.

Beach Plums

Many people, especially those from Cape Cod or the Jersey Shore, are familiar with beach plums, a native shoreline shrub of the northeastern United States. The fruit is small with a large pit-to-flesh ratio. Most people know them from the superb jelly they make. We've found them to be cold hardy and prolific on our farm. The shrubs grow eight to ten feet tall (2.4–3 m) and get bushy and thick, creating a nice hedge. They can also be pruned and cut back to make smaller bushes or can be shaped as a tree. When we've grown them into hedges, we've found their prickly branches a bit harder to circumnavigate when picking.

Beach plums bloom in May, when their small white flowers are filled with the buzz of native pollinators that look particularly lovely against the dark branches. Two different seedlings are needed to cross-pollinate. The plums mature in late August. Since our plants are all seedlings, every bush is a unique individual. Their fruits can be astringent or sweet like commercial plums, and vary in color from yellow, to red, to dark purple. We like to mix them all together, juice

FIGURE 8.7. Beach plums range in colors from yellow to purple. These red ones are ready to pick.

them in our steam juicer so there's no need to pit, and then make our Wild Plum Cider with 20 percent beach plum juice and 80 percent apple cider. The tannins give the cider a great mouthfeel, and the stone fruit flavor comes through. We have also canned them whole in a sweet syrup as sugar plums for a tasty treat in the winter.

Propagating Uncommon Fruits

We propagate many of our berry bushes and offer them for sale in our on-farm nursery. The currants, gooseberries, and elderberries are propagated from new-growth hardwood cuttings collected in late winter from dormant bushes. We also use hardwood cuttings to propagate our willow and dogwood species. Selling hardwood cuttings and scion wood to people who want to grow out their own plants or graft trees has become an interesting new business venture for late winter and early spring. We make the cuttings in late February and March, while the plants are still dormant. For hardwood cuttings and scion wood, we look for healthy shoots from the previous growing season

FIGURE 8.8. Propagation beds are used for hardwood cuttings.

TABLE 8.3. Other uncommon fruit grown on the farm.

Name	Notes
Gogi Berry (*Lycium barbarum*)	Though we've only recently planted these bushes, they've already produced a few nutrient-rich berries. The plants grew well the first season but did not overwinter. They are self-fruitful and supposedly zone 4 hardy.
Hardy Kiwi (*Actinidia arguta*)	Both male and female plants are required to produce fruit. The Anna variety grew vigorously during the summer seasons, but died back several winters over eight winters. One winter when it did not die back, it produced a few flowers on one vine. We had to cut the vine to collect a honey bee swarm, however, so we didn't get any fruit. The following winter, the vines all died back again. We decided to give up this year and cut them down.
Juneberry (*Amelanchier canadensis*)	Native to Vermont, this shrubby tree is also called shadbush and serviceberry. These are the first white-flowering trees to bloom on the edges of woods in our area. It bears small blue berries in June that birds love.
Juneberry / Saskatoon berry / (*Amelanchier alnifolia*)	This *Amelanchier* species has been cultivated to produce more and bigger berries. Used in commercial production primarily in Canada, they call these saskatoons. These look a bit like blueberries but are drier, with an apple-like taste and texture. Early blooming, June bearing.
Lingonberry (*Vaccinium vitis-idaea*)	Low evergreen bush about 10″ (25 cm) that flowers in summer and fall. It produces small, red, tart berries in fall. Winter hardy (zone 2–7). Loved in Scandinavia, this plant likes low pH soils with good moisture. Does well in partial shade. Self-fruitful but better pollination with multiple varieties.
Nanking Cherries (*Prunus tomentosa*)	A well-known bush cherry that produces small cherries in late July that are tasty and sweet when fully ripe. Birds often beat us to them in lean years. Fast growing and hardy to zone 3. Great for borders and edges. Two plants are needed for cross-pollination.
Seaberries, Sea Buckthorn (*Hippophae rhamnoides*)	Male and female plants required for pollination. Hardy to zone 3. The fruit is an orange citrus-tasting berry that ripens late fall. Our bushes have been relatively slow growing over the past few years, producing only a small amount of fruit. We think our soil may be a bit dry for them. Likes a neutral pH soil.
Wintergreen (*Gaultheria procumbens*)	Slow-growing evergreen ground cover, these produce strong, wintergreen-flavored berries even when young. Likes cool summers, hardy to zone 3.

that have multiple buds along the twig. We cut the currants about eight inches (20 cm) long and the elderberries so that there are two sets of buds on each stick. These are labeled and placed in ziplock plastic bags with a moist nonbleached paper towel. We store the bags in refrigerators until we are ready to plant.

Every year we fill multiple four-by-eight-foot (1.2 × 2.4 m) propagation beds—whose frames we built from old boards the previous owners had stored in the barn—with compost and soil. These propagation beds are situated near the nursery area, which allows us easy access for watering and weeding as needed.

In April when the soil has thawed and has started warming up, we plant the hardwood cuttings (currants, gooseberries, and elderberries) into the prepared beds and label them accordingly. The ideal conditions for root formation are warm soil and cool ambient air temperatures. Preparing the bed often involves weeding, repairing the frames, and adding more compost. When we push the twigs into the soil, we make sure there are buds aboveground. (These are where the leaves will emerge.) We tamp down the soil around the cutting to ensure good contact between the soil and the twig. In the row, we generally plant around ten inches (25 cm) apart for *Ribes* and a foot or more (3.7 m) for elderberry twigs. Between the rows, we usually allow twelve to eighteen inches (30.5–46 cm). We water these in to ensure good soil-cutting contact.

We also use tree tubes for propagation. These plastic tubes are two inches (5 cm) in diameter and ten inches (25 cm) long with drainage holes. We fill these with our potting mix and then push in the sticks, watering in well after planting. These tubes are great space savers, but when the plants are well rooted, they should be repotted into larger pots or planted in the ground in order to get them well established. Late summer can be tough on these new bushes in the tubes; they've rooted fairly well by then, so providing enough water on hot days can be tricky. The plants don't overwinter well in these tubes either.

For most of the species we propagate, we haven't found the need to use a rooting hormone, but dry soil and hot weather can negatively impact rooting. Also, if the soil is too wet from too much rain or from overwatering the tubes, the twigs can rot and not root at all.

We've propagated our aronia plants primarily by digging up suckers that have already established roots around the base of well-established three-to-five-year-old plants. We dig these up in the spring and pot them or grow them out if needed. We've recently heard that hardwood cuttings may work for aronia propagation. We will be trying it out this year. Beach plums can be grown from seed. We have identified a few individual beach plum plants with especially good fruit that we are grafting to propagate.

Honeyberries require softwood propagation techniques, which we avoid because of our workload in the summer. We generally buy in blueberry and honeyberry plants from wholesale nurseries in the Midwest. Unfortunately, these are not grown organically, so until those plants have been in our care for a year, they are not considered organic.

We also buy in small plants from nurseries and grow them out in designated nursery areas on the farm for a year or two. Many of our conservation plants are handled that way. This practice allows us to sell larger plants that are also organic, which is especially important when selling pollinator shrubs and trees. Many wholesale and commercial nurseries use systemic pesticides, which can remain in the plants and be passed on to pollinators via nectar or pollen. It is important to ask.

FIGURE 9.1. A honey bee visits a spring willow flower.

CHAPTER 9

a walk on the wild side
the pollinator sanctuary

Our conversion of the fourteen-acre back pasture to a pollinator sanctuary didn't happen overnight or even over a couple years. It's been going on for over ten years, and it's still a work in progress. We stopped grazing and mowing one part of the back meadow, and after a few years the milkweed, goldenrod, and aster came back naturally. We reclaimed the wetter, reed canary grass–dominated areas by planting willows, silver maples, American basswood and littleleaf linden trees, elderberry, and wetland roses. A few years later, we started planting apple trees on the south-facing slope of a silt knoll and continue to add more plants, creating our Knoll Orchard (polyculture apple orchard) and, later, Pear Corner Orchard. We started a willow labyrinth, which we finished planting in 2018. Recently, we added no-till perennials and annuals. We've also sold plant materials to florists and wreath-makers, and milkweed pods for seeds and floss, and grown native grasses.

We keep a pathway mowed around the perimeter of the pollinator sanctuary that we call "the pollinator pathway." This path allows us, our dogs, and our visitors to get up close and personal with the plants, insects, birds, and other wildlife that call the pollinator sanctuary their home as well as the ones that wander in from nearby.

Most of the shrubs, non-fruit trees, and perennial wildflowers we've planted in the pollinator sanctuary are native Vermont plants. It's important to plant natives when trying to increase wildlife biodiversity. Native plants are hosts to native insects including hundreds of moth and butterfly caterpillars, beetles, and bugs. Leaf-eating insects that have been demonized by human society perform an important ecological role. They turn leaves and plants into more meaty insects, which are in turn food for birds and other wildlife. In his book *Bringing Nature Home*, Doug Tallamy writes that "A plant that has fed nothing has not done its job." We take that seriously. Insects are the most important animals to turn plant material into animal protein, which is then used throughout the food web. Many of the ornamental landscaping plant species are not helpful when trying to enhance animal biodiversity, because native insects are not adapted to live on them. Gardeners often like this insect-free aspect of non-native ornamentals, but we hope that people's aesthetics will one day change so that they celebrate holes in leaves for what they represent: life!

The Pepinyè Garden

Our usual first stop on a tour or walk is our demonstration and nursery garden that we call the pepinyè garden—a Creole term that Haitian farmers use for a plant nursery. We decided on the name to honor the Haitian farmers John worked with in the Dominican Republic and Haiti. Our pepinyè garden started as a nursery for plant propagation, but we have since let many plants grow to full size and now use it primarily for a demonstration and riparian zone garden.

At any time during the growing season, there is always something blooming in the garden, thus providing floral resources for bees and other insects. The first plants to bloom are the willows in April, and the last are the witch hazels in October and November. Home to about fifty different fruit and pollinator shrub species (not including the various weeds) within 2,400 square feet (223 sq m), our pepinyè garden is a great way to demonstrate biodiversity. It also allows our nursery customers to see full-size plants and imagine how they might fit into their landscape as a hedge or specimen plant. They can't help but notice all the life-forms that depend on them. The birds here are amazing: the usual summer residents of catbirds, robins, warblers, juncos, and our favorite, the brown thrasher. In the spring, the male thrasher can be heard singing his heart out, often mimicking the songs of his neighbors. That these birds are also great insect foragers, scavenging among the leaf litter and soils, is a reasonable trade-off for the occasional berries they eat.

The patch of land that now comprises the pepinyè garden used to be part of the driveway and lawn up until the early 1990s. Its soil consists of a gravelly loam, and its proximity to the stream means it has good moisture content and fertility. It also floods occasionally, so we've selected plants such as buttonbush, elderberry, and aronia, which can withstand an occasional flood. It's a short-rowed jungle of shrubs and small trees such as American plum, nannyberry, and dogwoods, as well as perennial plantings of oregano, black cohosh, and blazing star. Things grow fast and lush, including the weeds, but one good weeding during early summer coupled with cardboard and woodchips is about all that's needed. Once the bushes and trees have fully leafed out, they help shade out the weeds.

The Streamco willow, a cultivar of purple osier willow, is the first shrub in the pepinyè garden to bloom in the spring. On a warm sunny day in April, it's loaded with honey bees gathering pollen for their brood. When the native bees emerge at the end of April, they can be found gathering pollen too. This species of willow is a favorite of basket-makers and was brought over by the early colonists for that reason. Although it's not a native, it still serves as a host plant to many different insects because it's a close relative to natives. We've made wreaths and a living hedge from it, but it can also be used for bank-reclamation projects and erosion control. This willow doesn't put out root suckers as other willows do, and because it's a male clone, it won't spread and compete with our natives.

In early May, the delicate sweet-smelling blossoms of the native American plum beckon native bees, flies, honey bees, and us. We often stand there for a long time, immersed in the fragrance, watching the activity of bees, flies, and beetles buzzing within the blossoms. The plums are ready to harvest in August. We've made a wild plum syrup from it that's sweet and tart at the same time. The American plum is listed as a threatened plant in Vermont. We like to encourage others to grow this wonderful native shrubby tree.

The pepinyè garden is a favorite haunt of the catbirds that return to the farm in the spring. These sleek gray birds can sound like a cat's meow, but more often than not they chatter at us to keep away from their nests. They try to sneak under the bird netting on the berries when we're not looking, but they also eat different types of arthropods

FIGURE 9.2. Buttonbush flower cluster.

and insects, including beetles, true bugs, caterpillars, and spiders.

No sooner have the plum trees dropped their petals than the currants, red osier dogwoods, and other species come into flower. Highbush cranberry shrubs are one of our favorites. They make a great hedge, as they are bushy and grow to about twelve feet (3.7 m) tall. They bloom in early to mid-June, and their bright red or orange cranberries stand out in late summer, fall, winter, and often the following spring. Though they look delicious and are technically edible to humans, their flavor reminds us of cat pee, unfortunately. The bright berries hang on the branches until the cedar waxwings come back in the spring and eat them all within a day or two. One day we walk by, admiring the red berries on the bushes that are starting to leaf out with green trilobed leaves, and the next day, all the berries are gone. After a long migration, those cedar waxwings are hungry.

In late June and early July, the native Virginia rose starts blooming. You don't need to stop to smell them, because just walking by gives you that rosy scent. Ours are about five feet (1.5 m) tall, of numerous upright canes, and planted close together in a thicket that could be used as a living fence, complete with thorns, if so desired. Tiny native bees as well as bumble bees gather nectar and pollen from the pink flowers. Often, we see semicircles or moon shapes cut from the leaves, a telltale sign of leaf-cutting bees. They line their nests with rose leaves like cigar rollers.

We think a pretty winter pepinyè plant is the Virginia rose, whose orange rose hips hold on all winter and on a frosty morning look stunning against the rusty canes. We've planted yellow-twig and red osier dogwoods nearby, and the combination of orange and red and yellow stirs the artistic soul.

Silky dogwood, buttonbush, and elderberry bushes are in full flower in July. These six-to-twelve-foot (1.8–3.7 m) native shrubs tend to grow in thickets in moist or heavy soils. They make great hedges and windbreaks and can also be used for bank stabilization. The creamy white flower clusters of buttonbush attract a range of flying insects. The long protruding styles extending from individual flowers make its white ball-shaped clusters look like flowers from an alien world. When the petals fall off, the remaining green bumpy balls mature into brown balls of nutlike seeds that are a favorite of ducks and other wildlife. The silky dogwoods put out the prettiest blue berries in August that can disappear before you get a chance to appreciate them, as they're another favorite berry of many resident birds and squirrels.

The panicles of the white elderflowers send out a dizzying scent in July. Elderflowers are wind pollinated, but honey bees and other insects also use them as a nectar and pollen source. Birds love the berries. We often harvest the edible delicate-looking flower heads in July and late-bloomers in August for making elderflower cordial. The flowers shed their pollen at this time, which can lead to sneezing and watery eyes if you happen to be allergic to the pollen. The scent is a little intoxicating, too, and we have felt woozy after harvesting sessions.

August in the pepinyè garden is the time for perennial flowers such as sneezeweed, black cohosh, and a few goldenrod and common burdock around the edges that have crept into the garden when no one was looking. Bees and butterflies, wasps and flies, are out in force, sipping nectar from these native attractions. Finally, in October, the American witch hazel blooms after all its leaves have dropped to the

FIGURE 9.3. Elderflowers in bloom create a beautiful native edible plant for landscaping.

ground. The scraggly yellow flowers with their spicy fragrance are pollinated by owlet moths as they feed on the witch hazel nectar. We've never actually seen these moths, but we think about them on cold nights when the temperatures drop to near freezing. They shiver to warm their bodies as they feed on the sweet nectar. Witch hazel is also a favorite tree of herbalists and plant medicine people because of the medicinal properties of its bark.

The lesson of the pepinyè garden is to create biodiverse plant spaces wherever you can and to marvel at the insects, birds, and other wildlife that move in. One of our favorite comments from a recent visitor was, "Wow! This place has changed. It looks like a jungle in here. You've created your own ecosystem." We think this is what all farms and homes should do: create their own biodiverse native "jungle."

Riparian Zones and Wildlife Corridors

Moving past the pepinyè garden, we enter the gap through the riparian zone, wide enough for the truck and tractor to get through when needed. While many of the trees and plants in the riparian zone grew up of their own accord, we've planted others too. Dogwoods (red osier and silky), black cherry, American plums, and black locust trees are a few of our favorites.

FIGURE 9.4. An old piano creates three-dimensional wildlife habitat.

We've kept the riparian zone contiguous with the pepinyè garden to create a wildlife corridor that connects it to the woods behind the farm. Continuity of food and shelter is important for wildlife. This biodiverse wooded corridor also creates an edge effect on both sides with adjacent open meadow or cropland, another important aspect for enhancing biodiversity.

We check the water level in the stream as we pass over. The stream usually dries up in August, but in a particularly dry year, it can dry up earlier. Our boys and their cousins used to try to collect the minnows and small fish that became trapped in the last remaining pool. They'd scoop them into buckets, and we'd take them across the road to the Lamoille River. That instinct to save a stranded animal is strong in little kids. It's good to nurture and cultivate that instinct in children because it seems that all too quickly (usually by middle school), it, too, dries up. Other things become more important than the fish in the stream.

The fish don't get scooped up into buckets anymore. These days as the stream dries up, birds such as our resident American bittern or a visiting great blue heron, raccoons, or maybe even our barn cats catch and eat the fish, or maybe they decay in the mud, food for bacteria and other microorganisms.

The relationships and processes within our farm ecosystem, and any ecosystem, are time and space dependent. Things change. The plant and insect communities in our fields and gardens differ between spring, summer, and fall and between the wet areas and the dry. Now that we've let the trees and shrubs come back along the stream, the riparian zone is nothing like the grassy stream bank it once was. The addition of trees and shrubs has had huge benefits in terms of increasing the bird population; we now have dozens of resident species where there was once only a handful.

Sometimes the addition or loss of one key species can have dramatic effects. When European settlers came to this part of Vermont in the 1790s, they heavily trapped the beavers, which in turn changed the course of the rivers and the shapes of the wetlands. They chopped down 80 percent of the forests in Vermont and turned the landscape into pasture for cows, horses, sheep, and mules. A hundred years ago the forests started coming back. Similarly, twenty-five years ago, our farm fields were mowed and planted right up to the stream they called a ditch. Now there are willows, birch trees, and a pepinyè garden in the riparian zone of the stream. Ecosystems are constantly changing.

And, yes, if you come to visit us, that is a piano off to the left. All our children learned to play on that turn-of-the-century, old, upright behemoth. The 1920s were the golden age of piano building, and the country is still flush with many of these models. We bought it for a song, and now after four tunings over the years, ours was beyond hope of retuning without having it totally restrung. One day when a group of students with strong backs was visiting, we hit on the idea of getting the piano out of the living room and repurposing it as three-dimensional wildlife habitat. Of course, the students were up for this reinterpretation and agreed to help move it. They also took turns playing it outside. Later, we drilled various-sized holes in it and began to watch it return to nature. After a year, it still looks like a piano, but the walnut veneer has delaminated and the ivory has fallen off the keys. We were especially relieved to see the ivory return to the earth without a human value put on it. Already we have seen chipmunks, wasps, and bees entering and exiting the piano. We even found the shed skin of a snake on the keyboard. Yes, that is a piano, now being played as habitat.

Stately Trees

Back around the start of the twentieth century, our farm was called Elm Grove Farm. When we scraped the barn to paint it a few years ago, we could see the remnants of the name. But when we moved to the property in 1992, only two large American elms and a few smaller ones were left on the farm. One stately elm tree in the backyard was at least a hundred years old and provided lots of shade for the family. We set up a swing on a branch for the kids. We had picnics and, one year, a Northeast Organic Farmers Association (NOFA) workshop under it. At the workshop, we brought out our Guernsey cow, Honey, to share our family cow experiences while everyone enjoyed the shade of that big tree.

The other old elm on the farm lived out back near Pear Corner. One year it started to look sick, its leaves yellowed prematurely; the next it was dead. A year later, the one in the backyard went the same way. Back in the early 1900s until the middle of that century, elms were everywhere in the United States. As kids, we remember those big beautiful trees in our neighborhoods. By the mid-1960s, they were almost all gone, victims of Dutch elm disease, an invasive fungus from Europe that wiped out about 90 percent of the elm trees in the Northeast United States. The fungus is carried from tree to tree by elm bark beetles. Our two trees survived for a long time on this out-of-the-way Vermont farm, only to finally succumb in the new millennium and on our watch.

We left the dead elm near Pear Corner as a snag for wildlife. A few winters later, a windstorm knocked it down. A snag's a dangerous thing when it's too close to home, so we hired arborists to take down the dead elm in the backyard and cut most of it for firewood. We had the big trunk log sawed into planks at our local sawmill. A carpenter made a nice picnic table  and two benches that now grace our nursery with a few of those planks. They remind us of that big elm we loved.

We couldn't decide what to plant near the old elm stump to replace it in the backyard, so we ended up planting three trees; a burr oak, a sugar maple, and a basswood. The new trees are growing too close together for all to become full-grown trees, which means in the future, when they start to compete with one another, we, or our successors, will have to decide which one to keep. For now, we enjoy watching them grow.

The burr oak belongs to the white oak family. While its natural range is mostly in the Central and Northern Great Plains, it's also found in eastern and even northeastern areas. It's a tall tree that grows up to one hundred feet (30.5 m) tall with trunks that can reach ten feet (3 m) in diameter. A host to a variety of insects, the burr oak also supports wildlife including squirrels, deer, bears, turkeys, and more with its acorns. Because they are low in tannin, burr oak acorns are also of interest to us as human food, so we've planted another burr oak out along the fencerow in the pollinator sanctuary.

No home in Vermont should be without at least one big old sugar maple tree on the property, so we planted one near the elm stump. It is the Vermont State Tree, after all, and the source of all that maple syrup goodness Vermont is known for. Our neighbors' sugarbush on the hillside behind the farm is a mix of sugar and red maples, and always attracts the leaf peepers with their cameras in the fall. We prefer the subtler pastel pinks and purples of their buds announcing spring.

Nancy grew up with a sugar maple in her front yard, which created a favorite shady place to play. And when she was eight, her mom dug up a couple from the nearby woods and planted them in the backyard after the loss of their elms. On the farm, we've planted another sugar maple out along the back fence in the pollinator sanctuary, for more food and habitat for insects and birds. With climate change, the Northeast hardwood forest association of birch, beech, and maple trees may become a thing of the past. With warming winters, red oaks and even the white oaks will extend their range northward, outcompeting the maples. For now, we are grateful for the maples, their beautiful colors in fall, and of course, the maple syrup.

FIGURE 9.5. Locust in bloom in early summer. *Photograph courtesy of Alisha Utter.*

The American basswood is a fast-growing native tree found throughout the Northeast and often valued as an ornamental shade tree. We love it because its flowers provide nectar and pollen for bees and other insects in late June in our area, and it's also a host to many native moth and butterfly caterpillars. Some people don't like it, because it's also a favorite of Japanese beetles that eat and often skeletonize the leaves. Its small round fruits provide food for many small mammals.

Since the demise of the big elms on the farm, we've taken more notice of the younger ones we've found in various places. One in the backyard, a daughter of the old one that died, is already thirty feet (9.1 m) tall. Others that were small when we moved in are looking like good-sized trees now. We've also planted several Dutch elm disease–resistant Valley Forge elm cultivars from Pennsylvania along the entrance to and throughout the pollinator sanctuary. Maybe future owners of this place will want to name it Elm Grove Farm again.

Black locusts are another one of our favorite trees. They're a pioneering species able to grow in impoverished soils. Great bee forage trees with edible pealike flowers, they also produce rot-resistant poles and dense firewood with high BTUs. You can also coppice them for continuous production. Plus, they fix nitrogen. Because of their propensity to spread, they are not loved by everyone. Some states (including Maine, Massachusetts, and Connecticut) prohibit or restrict the importing, selling, or trading of these trees, although this isn't the case in Vermont.

We cut down two good-sized black locust trees in front of the house a few years after we moved in. We thought our reasons were good at the time: they blocked the front of the house, were difficult to mow around, and weren't very attractive. They're commanding trees, and when you see them in front of old farmhouses, that's all you see. They also prevented us from pushing back the driveway and building a stone retaining wall in front of the house. But the main reason was we didn't really appreciate them back then.

Those locusts gave us a lot of firewood for a couple years, though we needed to blend the wood in with other less dense logs when burning it. Otherwise, the fire would get too hot and could have damaged the woodstove. For days after cutting the trees down, we fought with the stumps before we were finally able to remove the main parts. But those black locusts didn't give up. For years their hidden underground roots kept sprouting up new trees in the lawn, under the spruce trees and near the house, until we finally let a couple saplings grow. Once we let them have their way, there was no more sprouting in other places.

They put their energy into keeping those two replacements alive and well. Those trees are thirty feet tall now, overtopping the white spruces that gave them shelter and maybe sharing nitrogen to repay the favor.

Besides their unwanted suckering, another drawback to black locust trees is that their bark, seeds, and foliage are toxic to both animals and humans. Chewing on bark-covered locust sticks can be dangerous for dogs and humans. When a black locust tree dies, livestock can be inadvertently exposed to root suckers that grow up in their pasture. Although some silvopasturists claim locust is fine for sheep and other livestock, it's definitely bad for horses. We learned this the hard way once, when our mare Nora reached under the fence to eat the bark of a black locust sapling. We found her shaking and twitching, so we called the vet. The vet gave her charcoal to absorb the toxins, and Nora recovered the next day. We were lucky that it was a small amount of bark, and that we got to her quickly. We've heard about other horse owners who weren't so lucky. We've been much more careful since then.

Despite their drawbacks, we still regret the loss of those original black locust trees now that we know more about these stately trees. We didn't appreciate their bloom and the shade they provided the house in the summer. We didn't appreciate their ability to grow in that poor sandy soil in front of the house. Over the years, we've been trying to make it up to them by planting dozens of new ones on the farm.

Reclaiming Reed Canary Grass Areas

As we pass the recently planted elm and black locust trees at the entrance to the pollinator sanctuary, we notice the patch of wetland roses off to the left. The wetland rose blooms a couple of weeks after the Virginia rose, thus perpetuating that sweet rosy scent on the farm. Japanese beetles emerge from the soil in this area in the beginning of July, and because of the timing of the bloom, the wetland rose draws more of them than does the Virginia rose. And those beetles do like roses. We don't begrudge them, as there are plenty of roses to go around.

We've planted an elderberry patch and more highbush cranberry nearby. Looking past the roses, shrubby willow stands of pussy willow and black willow, nannyberry, and silver maple ten to fifteen feet (3–4.6 m) tall sway in the breeze. This is a wet place on the farm in spring and fall where once only reed canary grass grew. Reed canary grass favors wet and marshy areas and can easily outcompete other plants. It's often considered an invasive grass, although it's unclear if it was introduced to this area or is actually a native species. We planted the shrubs into the grass using woven landscape fabric to keep the grass from competing with the newly planted small trees and new-growth cuttings (in the case of willows and elderberry). The landscape fabric allowed the shrubs and trees to get established. After a few years, the plants were thick enough to shade out the grass, so we removed the fabric.

In the spring and early summer, we often see butterflies in this part of the sanctuary. It turns out that pussy willows aren't only a great early pollen source for hungry bees but also a host plant for viceroy and mourning cloak caterpillars, to mention just a couple. The pussy willows also make great cut stems for the March and April farmers market, when Vermonters' cabin fever has them grasping at even the tiniest hope of spring they see in the fuzzy gray buds of willows.

Nannyberry, another native shrubby tree, produces pretty white flower clusters in June and edible fruits in the fall. We've been growing them as a hedge, but they can also be grown as small trees. Nannyberry is host to the blue azure butterfly, one of the butterflies of spring. Imagine the simple pleasure of seeing a blue butterfly flitting among the unfurling green leaves in May.

Stands of silver maples make wonderful trees for wetter areas. They can grow up to a hundred feet (30.5 m) tall, bloom in early spring, and although mostly wind pollinated, can still be an early nectar and pollen source for bees. We planted ours to use for coppicing for firewood and ramial woodchips. A fast-growing tree, it sprouts back after coppicing and continues to provide habitat and food for insects and wildlife.

This reclaimed grass area, now filled with shrubs and small trees, connects with the woods that grew up close to the stream. By extending this riparian zone buffer into a multilevel hedgerow, we're providing greater biodiversity and habitat for wildlife. The variation in canopy height allows the birds, insects, and wildlife additional three-dimensional habitat.

Recently we've planted even more trees, shrubs, and perennials in this part of the sanctuary. Many are still small, but the ninebarks are already over five feet (1.5 m) tall. This native shrub has two common cultivars: green-leaved types with white blooms and orange seed heads, and maroon leaves with pinkish flowers and dark maroon seed heads. We tend to like green-leaved varieties of plants because we can't shake the idea that reddish, mottled white, or yellow leaves of some cultivars are a sign of pesticide or disease damage. It's also not clear if cultivars are as good at hosting insects as is the original.

European elderberry keeps trying to grow in our cold northern Vermont farm, but it's having a difficult time with the winters. We planted it over five years ago, and it keeps coming back in the spring even though it's still only a few feet tall. Plants really are amazing. This past summer, we actually picked a few panicles of big juicy berries from these hangers-on.

Several littleleaf linden trees keep trying to grow in this part of the pollinator sanctuary as well. We planted these years ago, and every winter the deer munch on the new growth. After about eight years, two are finally getting tall enough to beat the deer. Littleleaf linden isn't native to Vermont, but because it's related to the native American basswood, which we have in abundance out near Pear Corner, it also serves as a wonderful host plant to a variety of native insects. John enjoys munching on the small leaves in the spring as a tonic. He says the slimy texture reminds him of okra. It's a great tree for bees too.

Patches of swamp milkweed, New York ironweed, and black-eyed Susan along the path provide additional summer beauty for insects and for us. Swamp milkweed flowers are favorites of a couple of different kinds of spider wasps. These two-inch-long wasps—one a dark blue-black species, the other red and brown—dominate the patch with only an occasional bee or butterfly daring to visit.

Other native plants such as common hackberry, crab apples, and witch hazel are found on either side of the path. Hackberry has gained renewed interest as a shade tree in urban areas because it can withstand dry conditions. We wanted to grow it because it supports all sorts

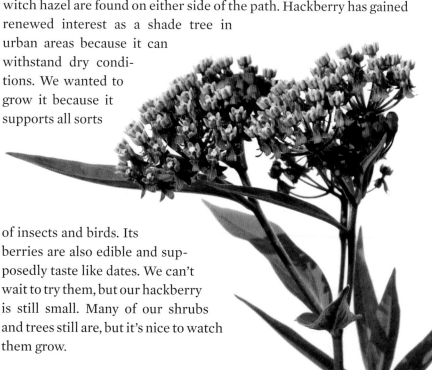

of insects and birds. Its berries are also edible and supposedly taste like dates. We can't wait to try them, but our hackberry is still small. Many of our shrubs and trees still are, but it's nice to watch them grow.

Willow Labyrinth

Labyrinths have been used for centuries to provide a meditative spiritual experience for visitors. Unlike a maze, which is designed to trick people into losing their way, labyrinths are designed to help people find their way, their spiritual path. The labyrinth can be walked while meditating (whereby you get out of your head and into your feet) or as contemplation (whereby you stay in your head and work out things that are troubling you). Besides all the ecological services the willow labyrinth provides, helping us and visitors on our spiritual paths was the major reason we included it in the pollinator sanctuary.

Long before we started planting the outer labyrinth circles, we planted swamp white oaks in what we planned to be the center or heart of the labyrinth. This oak can tolerate moist soils. Like the burr oak, it falls into the white oak taxonomic group and is a wonderful host for insects and wildlife. We're too far north to grow white oaks, so we opted for this close relative. Oaks were sacred trees for the ancient peoples of Ireland. Since John believes he is descended from the Irish kings, we thought this was a good way to honor his heritage.

Around the oaks grows a ring of winterberry holly. In Norse and Celtic lore, evergreen hollies were used to decorate the home in winter. Although winterberry is a deciduous holly (it loses leaves in the fall), the berries last all winter and are a favorite among the floral industry. This tall shrub is often used for hedges and borders. The species is dioecious, meaning individual plants are either male or female. Only the female plants produce berries, which are favorites with the robins and cedar waxwings and, for us, gorgeous highlights of the fall and winter.

We started the labyrinth in 2012 with these central plantings. Gradually, different species of willow have been added in seven concentric rings around the center, planted six feet (1.8 m) apart. We started many of these by planting willow sticks. We also propagated willows in tree tubes to let them root out a bit before planting. In still other cases, we planted shrubs grown in small pots, which required a bit more work, but these were extras from the nursery and we wanted to use them up. Mostly we planted hundreds of sticks.

One of the great things about labyrinths, unlike mazes, is that you don't need fully grown shrubs to walk it. In fact, many labyrinths are made of stones or patterns on the ground that you follow slowly. One day our willows will provide a hedge and a sense of privacy, but for now the labyrinth already serves its multiple purposes as a place for mindful walking and meditation while helping to reclaim areas of reed canary grass as part of a biodiverse planting.

Milkweed and Monarch Butterflies

In several of Knoll Orchard's alleys, thick stands of common milkweed grow as an alley crop. We found a way to favor milkweed in these alleys by grazing or mowing the mixed meadow early in the season when the grass is growing strong, but before the milkweed emerges. This gives the milkweed a competitive advantage. The pretty pink flower heads smell glorious in June and July, but even more impressive are the multitude of bees and butterflies that feed on the nectar. The buzzing hum and sweet scent draw our attention as we walk the pathway. Why wasn't this plant named "milkflower" instead of milkweed?

While the milkweed flower evolved all kinds of attractive strategies to get insects to pollinate it, the milkweed plant itself is full of deterrent plant toxins, including its sticky milky latex designed to keep insects from feeding on the leaves. Yet a whole ecosystem has developed around milkweed that includes insects that have developed detoxifying methods. Many of these milkweed feeders tend to be orange and black or red and black, the same cryptic warning coloration as the most famous milkweed eater, the monarch butterfly.

Well before the monarch butterflies migrate back to Vermont, however, the milkweed plant hosts parties of insects: the milkweed leaf beetle, a roundish chubby-looking red and black beetle; the red milkweed beetle, a cylindrical red beetle, a member of the long-horned beetle family that features black spots and long antennae; small milkweed bugs, mostly black with an orange-red X on their backs; and the large milkweed bugs, a larger version that possesses similar coloration but lacks the X.

The monarch is the most famous of the milkweed eaters. We feel a special affinity for this migratory wonder. John even had the privilege to visit their overwintering sites in the fir trees of mountainous Michoacán, Mexico. Amazing. Monarchs return to our part of the United States in July, with one generation eventually returning to the overwintering sites in central Mexico.

Monarch caterpillars feed exclusively on milkweed species. The loss of milkweed habitat throughout their migratory path is a major cause of the decline of the monarchs. This habitat decline is primarily due to the expansion of monoculture farming into meadows and other milkweed-populated areas and the broad-spectrum herbicides that have reduced the amount of weeds in and around fields. While it's easy to point fingers and blame monarch decline on one factor, others include increased use of new systemic pesticides, climate change, and ecological damage to overwintering sites in Mexico.

One of the basic attributes of any system, including ecosystems, is that everything is connected. These connections, or interrelationships, among species are often complex and not well understood, and may take the form of predation, parasitism, or mutualism. Likewise, many species' relationships with the chemical (both natural and introduced chemicals, such as pesticides), physical, and energetic components of their environment are not yet known. Being open to learning and respecting these interrelationships is a first step toward changing attitudes. Also, it's important to recognize that what we know about holistic systems is only a tiny percentage of what we don't know. This applies to human social systems, Earth ecosystems, and solar systems.

FIGURE 9.6. A monarch caterpillar feeds on a milkweed plant in the orchard alley. *Photograph courtesy of Alisha Utter.*

The butterflies that return to us in July are the great-grandchildren of those that left Mexico earlier in the spring. Vermont is one of the northernmost and last spots on their migration path. We usually see a few flitting around the pollinator sanctuary in July, sipping nectar from milkweed and other wildflowers. We also see the famous monarch mimic, the smaller viceroy butterfly with its more erratic flight pattern and black crescent on its hindwing.

At the end of July and in early August, we start scanning the milkweed plants for larvae. It can be hard to find the monarch caterpillars since they blend in so well, but it's always exciting when we see them. We continue to visit them over the coming weeks to watch their progress. We also see hairy and tufted milkweed tussock moth caterpillars on the underside of the leaves munching steadily. The tussock moth caterpillars (which we affectionately call "tussies") have a similar black and orange coloration as many of the species that feed on milkweed, although their orange is more subdued and coppery. Their fuzziness is made up of irritant hairs, so we resist the urge to pet them.

From late August to late September, if we're lucky, we are able to witness the pristine newly emerged adult monarch butterflies. Their bright orange colors stand out against the green and yellow background of late summer. They are not tattered and worn like their parents were. They sip nectar from the joe-pye weed, goldenrod, and flat-topped aster in bloom. The earliest ones to emerge may mate and produce another generation in our back meadow. Those butterflies that appear in the later part of the season will make their long migration back to the Mexican overwintering forests after fattening up on the nectar from the farm. For people who feel inclined to mow their meadows and fields, it's important to wait until October to mow. Not only do we want to make sure all caterpillars have a chance to turn into adults, but we also want to leave late asters and other wildflowers for the bumble bees and other bees that still need the nectar and pollen supplies. It's little things like this that can make a difference if we all do them.

The fluffy white part attached to the milkweed seed, known as milkweed floss, has gained new momentum in recent years as a substitute for goose down in comforters and clothing. It's hypoallergenic, insulating, and buoyant even when wet, and unlike down, it doesn't come from tortured animals. One year, we harvested the milkweed pods and sold them to a company in Canada that used the floss to make clothing for the Canadian coast guard. Though they're becoming more popular, these applications aren't a new idea. Milkweed floss was used during WWII for life vests to transport US troops, when the Southeast

FIGURE 9.7. A milkweed seedpod has opened.

Asian kapok tree, which previously provided the stuffing for life vests, became inaccessible due to Japanese occupation of the region. In the United States, civilians (including schoolchildren) chipped in by harvesting the pods and selling bushel baskets of them for a nickel. Our parents' generation remembers this.

We've also harvested milkweed seeds from our fields and sold clean seeds to farmers and homeowners for growing their own milkweed. The seeds can also be used to make oils, which garner hundreds of dollars per ounce to be used in the cosmetic industry. Growing milkweed in wild spaces is an easy and fascinating enterprise for gardeners and landowners. Probably the easiest way to plant them is to scratch a patch of soil in the fall and sprinkle the seeds around. Milkweed seeds, like those from many other native wildflowers, need cold stratification to germinate. Cold stratification means moistened seeds need to undergo a cold period before they can germinate. Another method for cold stratification involves placing the seeds in a moistened nonbleached paper towel within a plastic bag and storing in the refrigerator for a month or two before planting in the spring.

Open Spaces and Meadows

Within the pollinator sanctuary, we've kept separate areas open for grazing the horses, for grassland bird habitat, and for wildflower meadows. Our horses graze on pasture for about six months during the year, so those areas are important. But in keeping spaces open, we're also thinking about the farmers who will take over the farm in the future. We want to leave them flexibility in what they can do on this land. We hope they'll take advantage of all the fruit plantings, but they may want to graze sheep or other animals as part of their farm operation. And maybe, just maybe, we'll get sheep again or grow vegetables in those areas. We're not that old, after all.

We continue to use management intensive grazing in the grazing areas, and thus they're filled with good lush grass, clovers, and forbs without a lot of perennial shrubs and unpalatable wildflowers. We don't graze other areas at all, and depending on the soil type, they may remain grasslands—which are great for bobolinks, certain sparrows, and meadowlarks—or become wildflower meadows of goldenrod, milkweed, and asters.

There is no shortage of goldenrod in the pollinator sanctuary. It lines the pathway in many places, and a portion of the back meadow is full of this wonderful pollinator plant. We have at least four species on the farm that start blooming in late July and continue into early September. The bounty of bees, butterflies, birds, and other animals on these flowers inspires us daily, though it might also be a source of frustration for our neighbors, so many of whom feel compelled to mow their fields at this time of year. Unless they're making hay, we don't understand why they mow during the summer instead of waiting until after the bloom. Maybe they like green instead of a sunny yellow. We've met honey bee keepers from the South who commented, "What we wouldn't give for a goldenrod field in August."

We feel the same way, except we're not only thinking about the benefits to honey bees. We're thinking about all the bees, butterflies, flies, and other insects that feast on this wonderful flower. In the late fall and winter, the seeds from these flowers will support chickadees, native sparrows, and an occasional flock of migratory snow buntings. Now, that's a sight. Nancy also uses goldenrod as a natural dye for wool in her fiber arts: a bright sunny yellow, like the flowers.

Native Grass Plantings

As we continue to walk our pollinator pathway, we pass a recent planting of native grasses, part of a project with the United States Fish and Wildlife Service that focused on growing these native grasses from locally sourced seeds, so selected because local seeds are genetically adapted to our environmental conditions. Native grasses serve important ecological functions, such as food and habitat for native wildlife, that are often not provided by alien grasses. Restoring native plants, including native grasses, is an important component of ecosystem regeneration.

Northeastern Pollinator Plants, a local nursery specializing in native plants important for pollinators and other wildlife, grew six native grasses (*Agrostis perennans, Calamagrostis canadensis, Carex gynandra, Danthonia spicata, Elymus virginicus,* and *Poa palustris*) from seed and provided them to us for the project. We used black landscape fabric with equally spaced holes to establish them in a pretty tough spot that was wet in the spring and, not surprisingly, filled with a thick stand of reed canary grass. Even with the landscape fabric, the reed canary grass crept in around the holes where the native grasses were planted. In a few cases, the canary grass choked out the new plants. In other cases, the native grasses gained a foothold and came back the second year. We've collected seeds with the hopes of reestablishing these native grasses in other areas on our farm.

Edges and More Trees

As we round the corner of the northern edge of our property, along the dilapidated old barbed wire fence woven between the shrubs and trees, a statue of a goddess sculpted by a local artist greets us and our visitors. She reminds us to reflect on and appreciate the feminine aspects of the divine, such as nurturing and fertility. If we're lucky, we might see a garter snake sunning herself nearby. Or if we stop here in the evening to pay our respects, we might hear the hooting of a barred owl up in the woods. We imagine her black eyes staring out into the approaching darkness hooting to her mate, "Who cooks for you, Who, Who?"

Hawthorn, black cherry, white ash, willow, and a variety of shrubs have grown up along the fence. Wild peppermint creeps into the field from the hill behind it and lines the pathway in July, refreshing to nibble on in the hot sun. In winter, chickadees and blue jays greet us from this boundary of our property. When we turn to look back the way we've come, we marvel at the mountains and the colors, no matter the season.

These edges make great homes for birds and other wildlife, so we've plopped down a chair under the shade of an ash tree next to our statue of Saint Francis of Assisi, the patron saint of animals and ecology and a nod to John's Catholic upbringing, and there we absorb the view and the sounds of birds and insects. We recently planted two mock orange shrubs and summersweet clethra, to enhance our experience as we take a rest. And nearby, we've added burr oak, willows, native roses, and buttonbush, just because.

FIGURE 9.8. The view from the back of the pollinator sanctuary is always a feast for the eyes, with Mount Mansfield in the distance.

The woods beside the stream that runs through our property provide another ecological niche. Big black cherry and ash trees keep the stream shaded. Their roots hold the stream banks in place during the spring thaw when the water runs fast and deep. Ostrich ferns (the species of fern that produces edible fiddleheads in the spring) grow lush and tall in summer, swaying in the cool breeze. This is where we see the most dragonflies: big green darners, common whitetail dragonflies, and red meadowhawks.

A new kestrel nesting box adorns the big ash tree that overlooks Pear Corner. Although we've seen kestrels hovering in the sky, looking for voles in our meadow, none have moved into the box yet. We are patient and determined and will try another approach, using a box mounted on a pole in a more open spot. The majesty of that big ash calls out to us as we walk by. Unfortunately, it might be gone sooner rather than later, as the emerald ash borer, an invasive beetle from Asia, has made its way from the Midwest into Vermont. This long-horned metallic beetle has already left millions of ash trees dead in its wake. Other diseases are also hurting ash trees in these times of climate stress. Change does not come without heartache.

By the big basswood where the boys built a rough tree fort twenty years ago, we've put in a small footbridge over the stream. For years, we waded through the water or balanced on the ice to make it across the stream at this spot. A gift for improvements to the pollinator sanctuary prompted us to make it easier for both us and our visitors to continue on the pathway. The boys' fort has since been reduced to splinters by the growth of the multiple basswood tree trunks it once spanned.

In this part of the sanctuary, basswood trees of all sizes prosper near the stream. We've seen an occasional porcupine here as well. Porcupines like to eat basswood bark and often live in rock outcroppings, or ledges, where they find snug holes to make their home. There are a lot of such ledges in the neighboring woods. Pulling quills from our dog Scout's nose has been a drawback of having porcupine neighbors.

Stone Fences and Vernal Pools

As we head back toward the production fields and the house, an old stone fence runs along the east side of the property. On the other side of it, in our neighbor's woods, overgrown apple trees, the remains of a nineteeth-century orchard, intermingle with basswood, gray birch, serviceberry, and other tree species that have grown up in the decades since. This stone fence was once part of a system of paddocks used to fence in the mules raised on this land during the Civil War. The owner of our farm and the adjacent farm, Chauncey Warner, supposedly made a fortune supplying mules to the Union Army during the course of the war.

The edges between forest and meadow are diverse in plants and wildlife. Blackberry and wild raspberry grow along the stone fence, trying to creep into the meadow or grab us as we walk by. Blackberry thorns hurt. Occasional mowing of the pollinator pathway cuts them off from the rest of the field and us. These plants, though, provide good pollen and nectar resources for the bees in June and berries for the birds in July and August.

As we walk along the edge of the woods, we listen to birdsong and try to discern the singers. Song sparrow, yellow warbler, common yellowthroat, hermit thrush, and ovenbird are the usual spring chorus. We might also hear the northern oriole and on summer evenings the deep-throated *gunk-gunk* of the American bittern. The melodious mimicking songs of the brown thrasher always stump us for a minute as we try to figure out what bird is singing all the different songs in the treetop.

Fritillary butterflies flit among the joe-pye weed in August along this edge. Their large size and orange and black wing patterning might, at first glance, be confused for a monarch. A better look will put things straight. They're not quite so large nor so bright as the monarchs, and their coloration is distinctly their own: rusty orange and black with bluish dots underneath. While the adults sip the nectar from many different flowers, the caterpillars are specialists and eat violets. In the springtime, the overwintered larvae awake and search out new green leaves of violets, of which we have plenty, before they pupate and become adults.

John dug out a few shallow holes with the tractor bucket in August one year to create vernal pools for the coming spring. This area of the land has several natural seeps that keep it wet for most of the year. He scooped out a shallow trench for the inlet and bigger holes for the pools. Sure enough, in late spring and early summer, these pools are covered with thick gelatinous masses of frogs' eggs floating on the surface. Frogs poking their heads up in the slime often dive into the water before we get a picture, or even a good look. We're so lucky to live in a place with clean enough soil and water to harbor a diversity of amphibians. Whether it's the chorus of spring peepers in May and June, the red efts on the trail, or a beautiful northern leopard frog in the garden, they are all appreciated and welcome on the farm.

Rest

The final leg of the pollinator pathway takes us back through our fruit and nursery production fields. In the spring we note bumble bees working among the rows of black currants, even on cold cloudy spring mornings. And come summer, there's a bountiful crop of black currants thanks to their efforts. By September, field 3 looks wild and scruffy, our poster child for farming on the wild side. A mix of pears, wild plums, rhubarb, currants, and nursery crops grow alongside plots of late-blooming Maximilian sunflowers, a native perennial sunflower. Later in the fall, the sunflower seeds are a good source of food for seed-eating birds.

One of the most poignant times to walk the pollinator pathway is in late autumn after the red, orange, and yellow leaves have fallen from the trees and everything has turned to rusty browns, mauves, tans, and grays. The insects are gone, and except for the resident birds, most have migrated south. This is a time for rest, both for the land and for us. A time to contemplate the life cycles of all the inhabitants of this place.

FIGURE 9.9. An ant mound rests in the off-season. Even ants grow sluggish in winter.

CHAPTER 10

rethinking pests, invasive species, and other paradigms

As a farm that values diversity, we don't have many biological enemies. At least we don't think of them that way. Certain plants are allies in the places we want them, sequestering carbon, filtering water, and creating fertility, and become "weeds" only when they compete with our plantings. We do spend a fair amount of time and effort on weed management. For our crops to thrive, we need to keep the grasses and other weeds out of the propagation beds and berry bush plantings, and in check during the early years of fruit tree establishment. Weed management isn't always easy, though. Plants have a lot of life force and want to grow!

We use a variety of techniques to keep the weeds at bay, including hand weeding, landscape fabric, and sheet mulching with cardboard close to the plants. Landscape fabric is great because it can be reused year after year, but it's not a good idea to leave it down over the winter, because it provides a safe cover for voles, which can then be a problem in themselves by girdling trees and shrubs in the winter. There's a pesky rodent! If we use proper vegetation management around bushes and shrubs, including vole guards on the trees, we can keep their damage to a minimum. They then become just another coinhabitant of the land. Did we mention that we like foxes?

We put nets on certain crops such as honeyberries, red currants, and blueberries before they ripen, to protect them from birds that are "pests" for only a short time in the summer. We take the nets off when we're done with the harvest, allowing the birds to clean up the fruit. This strategy helps reduce other pest problems, like spotted wing drosophila, in our early berry crops.

Our diversity of crops allows us to withstand occasional pest outbreaks and potential crop losses by leaning on the income from other crops until a balance is reestablished. We depend mostly on biodiversity and ecological intensification practices to limit insect pests. We should differentiate here between pests and pest outbreaks. You sometimes hear or read in alternative agriculture circles that pests on a plant reflect an unhealthy plant or soil conditions. We don't think that is correct. We consider having a diversity of insects, including those that eat crop plants, as part of the natural balance. A pest outbreak occurs when the population of an insect increases to the point where it can cause high economic damage to a crop. A low level of pests keeps the predators and parasites fed and happy and working. Wiping everyone out with insecticides causes wild fluctuations in pest populations, as their populations tend to grow faster than predators' populations.

If we grew only one crop, with clean fields and field margins, we, too, would be anxious about pests and potential losses; so we can understand why monoculture farmers turn to pesticides for insurance and assurance. But it is a treadmill, and we don't understand why they keep insisting on monocultures. Of course, the chemical companies know how to exploit fear with their marketing. Many of the pesticide advertisements in the trade magazines look like horror movie billboards, with giant caterpillars and their gaping fangs coming to get you. Pesticide companies are not necessarily the farmer's friend. They are more like parasites themselves, making their living off farmers. A good parasite doesn't kill its host, though.

The goals of efficiency and economy of scale (which is purely about maximizing profits) are still part of the mainstream cultural mentality in our society. We've shown that a small-scale diverse farm has certain ecological and economic advantages. We try to be efficient in our labor and harvests, but it's not critical to our bottom line to wring every last berry from the land. We can afford to share a little with our coinhabitants. As regenerative farmers, we're trying to find the right balance for us and the so-called pests to coexist.

This isn't to say that it's all unicorns and rainbows at The Farm Between. Pest outbreaks occur because of things beyond our control—like the weather or newly introduced pests such as SWD. There's this idea floating around that if you have a biodiverse organic farm with good soils and good management that everything will be pest resistant and in perfect balance all the time. No way. Species populations are cyclic; outbreaks and subsequent crashes occur. That's often how an overall balance is maintained. It's good to remember. In the following sections, we've included a few examples of how we relate to "pests" on the farm.

Japanese Beetles and Wild Parsnip

Japanese beetles seem to be every gardener's archenemy. Yet if you can forget about everything you've heard and learned about them in the past, get past the holes in your rose leaves, their clumsy flying and propensity for getting tangled in your hair, and look at them with fresh eyes, they really are quite lovely. Their shiny green thoraxes and coppery red wing coverings shimmer in the sun like glass bulb ornaments on a Christmas tree. The little white tufts of hair around the edges of their abdomen look like the fringe on a holiday dress. And the very cool fingerlike protrusions on their antennae are amazing!

Japanese beetles are members of the scarab family (Scarabaeidae) of beetles, the group of beetles revered by ancient Egyptians. The insect order of beetles (Coleoptera) contains 25 percent of all known species on the planet, about four hundred thousand different species of beetles, and more are being discovered all the time. Japanese beetles originally came from Japan and were introduced into the United States in the early 1900s. They spread quickly throughout the East and continue their range expansion in the States and beyond. Their larvae, or grubs, live in the soil for about ten months, eating the roots of grasses, which is another good reason to minimize the size of lawns. The adults emerge in the beginning of July (in our part of Vermont) and rapidly congregate and feed on the leaves, flowers, and fruit of just about everything. One article we read mentioned they enjoy over three hundred species of plants in the United States. They also spend considerable time mating. Females lay their eggs in the soil, where new grubs emerge and start eating roots. They overwinter as grubs and pupate in the soil in June.

FIGURE 10.1. A Japanese beetle visits a wetland rose. Note the parasitic fly eggs on its thorax.

We've collected many basswood leaves skeletonized by Japanese beetles and pressed them in books. The lacey leaves look so beautiful. They'll be great in an art project one day. Some leaf damage is not usually a problem, as trees and shrubs can often handle a fair amount of defoliation. The exception might be if the trees and shrubs are young and newly planted because they might have only a few leaves to begin with.

In the early years, we used to shake adult beetles from our rosebushes or from young cherry and plum trees into a bucket of soapy water. They'd fall in and drown. It's probably been over ten years since we've done that. And since we've changed our attitude about them, we've realized we don't seem to have a Japanese beetle problem anymore. Part of the reason could be we've reduced the size of our lawns (fewer grubs), and we've created a biodiverse and abundant landscape. If there are only a couple of fruit trees within a vast lawn that have helped feed and grow countless beetle babies, then of course come summer, there will be a lot of beetles on those trees. After reducing our lawns and planting hundreds of plants around the farm, we find that the beetles seem to be fewer in number and more spread out. The increase in biodiversity has also helped increase predators, parasites, and competition, which has likely impacted the Japanese beetle population.

We've also noticed more Japanese beetle parasites in recent years, especially in early July, when the beetles first emerge from the soil. White dots cling to their thoraxes. These are the eggs of a tachinid fly, a Japanese beetle parasitoid introduced in the United States from Japan back in the 1920s. When the eggs hatch, the fly larvae, or maggots, burrow into the flight muscles of the beetle, killing it shortly thereafter.

Besides Japanese beetles, another invasive species that nobody seems to like is the wild parsnip, which is encroaching into our fields and spreading farther and farther from the road, where it was first introduced by the highway department's mowing machines. What's interesting is the tachinid fly that attacks Japanese beetles likes to drink the nectar of the parsnip flowers. The reason we have so many parasitized beetles may be because of this new food source for the flies.

Getting the juice of wild parsnip on your skin and then exposing the skin to sunlight causes a blistery rash, a form of phytophotodermatitis that can burn and leave scars if it's a bad case. One year, our sixteen-year-old son decided it would be cool to make scars on his arm from the parsnip. He took the juice and made a couple lines on his biceps. Sure enough. it blistered and eventually left scarring, but it finally faded after a few years. By then, he had moved on to tattoos.

The risk of phytophotodermatitis and the fact that parsnip spreads quite readily make it a noxious weed. We do pull it from our berry bush rows by hand. It spreads only by seed, so the key to preventing the spread is to cut down the flower heads before they produce seeds. This is often easier said than done. Although insects like the flowers, we haven't seen too many that eat the leaves, although that might be changing too. Recently on the parsnips around the farm, we've witnessed a type of caterpillar that makes webbing on the unformed seed heads to build its protective home. In such flower heads, few seeds will be formed. We leave parsnips where we've seen the caterpillars to encourage more of this type of parsnip-eating webworm. We've also seen an occasional black swallowtail larva on the parsnips. These plants get left alone as well.

Wild parsnip is another one of those species introduced into North America by the early European colonists. Either it escaped from their cultivated parsnips or they accidentally brought wild seed from Europe. It's been in North America for a long time, but it seems to be spreading more in our area over the last couple of decades due to human intervention. Unfortunately, it's one of the few plants that Japanese beetles don't eat.

The Most Invasive Species

Humans have been moving plants and animals around the world for millennia, on purpose and accidentally. That's what our species does. Beside parsnips, early European colonists introduced pigeons, dandelions, Queen Anne's lace, bishop's-weed, earthworms, honey bees, apples, clover, and the list goes on and on. All of these are ubiquitous on our farm, in the United States, and around the globe. When it comes to invasive species, though, *Homo sapiens* tops the list. Talk about displacing native species and causing a loss of biodiversity! We are spearheading a mass extinction event, and yet we get our boxers in a bunch over other "invasive species." We win as the species causing the most disruption to the planet these days. We are even getting our own geologic epoch named after us: Anthropocene. Maybe we should tone down the rhetoric about our newest uninvited inhabitants. We've created this recombinant ecosystem, after all.

We shouldn't be too hard on ourselves, though. All species manipulate their environment for their benefit. Humans are a classic case. We went from two billion to seven billion in a hundred years. This increased our genetic diversity and distribution. Like an algal bloom or the scum on the pond, we have reached the point where we have overpopulated and overextracted our resource base, also known as the carrying capacity. This density-dependent depletion of resources with its attendant increase in toxins, disease, climate change, and, in the case of humans, war eventually results in the collapse or right-sizing of the population.

Hunter and gatherer–type peoples are often heralded as living more in balance with their surroundings than settled populations. That's true. They probably did (and maybe still do), but they often depleted their local resource base and moved on. Polluting and depleting the resource base has been the way of our species. If we think about the Romans, they polluted their own drinking water, the Tiber, which is why they made aqueducts. They conquered, exploited, and moved on to get wealth and resources and enslave nearby peoples. The overpopulation of Scandinavian countries in the latter part of the first millennium and the lack of arable land for the next generations led to Viking conquests, exploration, and migration. Immigration to the United States in the late 1800s and early 1900s was due in part to an overpopulated Europe and its lack of opportunity for younger generations. The continued immigration from upheaval in Latin American countries into the United States today tells a similar story, that of people looking for refuge or opportunity. It's human ecology, and it is predictable. In a more thoughtful world, the tragedies that produce refugees would be preventable.

We know so little about how ecosystems function and the countless relationships and interactions within them. Humans have always knowingly and unknowingly changed the environment to the detriment of some species and the benefit of others. Poor land management practices often create opportunities for newly introduced plant and animal species to proliferate, yet we continue these same practices and then often turn to pesticides to deal with the problem of these "invasives." As Tao Orion notes in *Beyond the War on Invasive Species*, this type of restoration work does not solve the systemic problems that led to the proliferation of invasive species in the first place. Herbicides treat only the symptoms and can wreak havoc on an already disrupted ecosystem. Depending on the species, physically digging up or pulling a plant can cause it to spread even more.

Ecological thinking requires us to reexamine our strategies when dealing with what some call invasive and noxious species. While intentionally planting them isn't recommended when trying to increase biodiversity on the land, more positive approaches for dealing with unwanted species need to happen. The first thing might be to flip your thinking and appreciate any positive attributes about them. Maybe they are edible, beautiful, help stabilize stream banks, or are a good pollinator plant. All these characteristics are true of Japanese knotweed, for example; however, it is typically framed as nothing but bad. If you still feel compelled to hate and fight them, planting native species and creating niches where the native plants can outcompete the spread of invasive species might be a better way to spend money and time than the chemical warfare of herbicides. Nature abhors a vacuum, so creating and leaving a blank space in an ecosystem often continues the cycle of pest outbreaks, whether by insects or plants. It can be a slow process to fill niches, though, and may take years to get it "right," which in the natural world means a more stable ecosystem.

Spotted Wing Drosophila (SWD)

This little fruit fly, *Drosophila suzukii*, has become enemy number one of berry growers in much of the United States. Originally from Southeast Asia, it was first noticed in California in 2008 and has since spread throughout the United States. It made its way to Vermont in 2012. We first found it in our everbearing strawberries toward the end of June that very year. When we realized what was making the berries mushy, we panicked and tilled the whole patch under, hoping that would prevent the fly's spread to the rest of the farm. In hindsight, that was pretty silly. After all, you can't keep a pest like SWD out.

Unlike the regular old vinegar fruit fly that can only lay her eggs on rotting or damaged fruit, SWD females have a knifelike serrated ovipositor—the egg-laying part of the female—that can cut into unripe and undamaged fruit. After cutting her way inside, she typically lays one to three eggs. The males are easy to identify because of their spotted wings. The adults and possibly pupae overwinter in leaf litter, duff, and rotting fruit. During the growing season, adults typically live for a few weeks. A female can lay three hundred or more eggs in her lifetime. Depending on the temperature, it takes a week or two to go from egg to adult, but they can cycle through many generations per growing season if you have a variety of fruit, as we do. As the season progresses, their populations can increase to damaging levels.

We no longer grow everbearing strawberries, not only because of the SWD, but also because we want to stick with perennial fruit. We've never seen them in the honeyberries, another reason we love these early berries. They show up in the blueberries usually after a couple of weeks into the picking season. When our fall raspberries start ripening in August, it might be a week or two before we start noticing damage. They're in the elderberries come late August and September. We've even seen them in fall native fruits such as highbush cranberry and silky dogwood and wonder what the ecological ramifications are of reducing these food sources for wildlife.

SWD numbers and their damage seem to vary based on the weather. They don't like it hot and dry. Unfortunately, the weather is beyond our control. Damage also varies based on picking hygiene, and this is within our control. Picking often and picking clean is the best way to curb their numbers and their damage. This also means sorting berries during or after picking. We need to do this anyway for other pests such as slugs, snails, Japanese beetles, and birds. Usually, the impact of these other pests is minimal, however. SWD can be severe if we're not careful. The times we've been lackadaisical about picking regularly, or when other people pick who aren't thorough, are the times we might find considerable numbers of SWD-damaged berries.

We try never to leave mushy berries on the bush. The really damaged berries get solarized in clear plastic bags if the numbers are high, or in the case of raspberries, we smoosh them onto the black plastic walkways or in the dry duff in the greenhouse floor when there are only a few. If this doesn't kill them outright, it will cause the larvae to dry out and die. Lightly or slightly damaged berries can still be used to make fruit syrups. Berries with no visible damage are sold fresh or frozen. The raspberry receptacle that is left on the stem after picking will be white when there is no damage and stained pink if a "young one" (code for SWD maggot) has been feeding there.

Even though we don't like this new pest, we've gotten over it to some extent. We figured out how to manage and live with it. We accept that we're going to have losses, which might be as high as 20 percent in raspberries, given the weather and management conditions. With a wet summer, elderberry losses can be even higher. We've experimented with netting individual elderberry panicles using the type of nylon sock that shoe stores use for trying on shoes. Enclosing the

FIGURE 10.2. We've experimented with using nylon footies to exclude SWD. It works well but is labor intensive. The unprotected berries on the right have dropped early from SWD. *Photograph courtesy of Alisha Utter.*

panicle of green fruit within the nylon sock and closing it off at the stem with a twist tie prevents access to the fruit by SWD but allows the berries to ripen. Given the time involved to sock each panicle, we figure this makes economic sense only when SWD numbers are high and with big-berry, heavy-producing varieties like Marge.

A Bit about Apple Pests

Whole books have been written about apple pests and diseases, so we'll say only a few things about them here. The belief that we could grow apples only by using pesticides kept us from planting apple trees years ago. We finally overcame that notion with the idea that we would sort apples and make organic cider products with the gnarly fruit.

We decided to plant mostly disease-resistant varieties such as Freedom, Liberty, Haralson, and Gold Rush in the Front Lawn

FIGURE 10.3. Red-humped caterpillars, just before we knocked them off, can cause serious defoliation in a newly planted or young tree but likely would not be a problem in a mature tree.

Orchard. We've planted many heirloom and other varieties in our polyculture orchard, and many of these are disease resistant as well. The most common diseases for apples in our area are apple scab and fire blight. Weather influences the spread and severity of these diseases, but starting with disease-resistant cultivars (whether heirlooms or new releases from university breeding programs) is something we can control. We can also control soil fertility, weeds and grasses in the early years, and organic matter. We put our efforts into those.

Insects can also cause a fair amount of damage to apples. European apple sawfly and plum curculio enjoy the early fruitlets. Apple maggot is the larva of a fly that burrows little railroadlike tunnels into the apples. Codling moth caterpillars burrow into the core of the apple, where they pupate, and emerge from the apples the following spring. A range of caterpillars, beetles, and other insects eat the leaves of apples as well. Apples are loved by all.

Mature apple trees can generally withstand a certain amount of defoliation, so we don't tend to worry about these leaf-eaters. In young trees that don't have a lot of leaves, however, we keep an eye out and knock off caterpillars as needed.

Hygiene is one strategy we employ to deal with the apple-eating insects. Since many of these insects develop or overwinter in the apple drops, picking up drops and hot-composting or feeding them to animals is a good way to reduce these insect populations. Even when we thin apples by hand in the early summer to increase the size and quality of the remaining fruit, we select the damaged fruit, collect them in buckets, and hot-compost these little green apples.

Through our diverse plantings of crops and companion plants, we've generated natural beneficial insectaries and large populations of birds that help support our pest management approaches. We are committed to manipulating our farm environment to enhance predator populations. Besides biodiverse plantings, we are including nest boxes for birds and solitary insect-hunting wasps and even earwig shelters within our orchards. One hundred years ago, most rural Vermonters had apple trees on their homesteads and used the apples for cooking, fresh eating, and hard cider. There were arsenic-based insecticides and sulfur fungicides available then, but most people weren't using them. And they enjoyed both great healthy fruit and healthy insect and wildlife populations. So what if their apples got a little gnarly? The idea that we douse our foods with poisons to enjoy them is outdated and outlandish. It's a mind-set that changed within a couple generations in the twentieth century. And it can change again.

Gooseberry Sawfly

The gooseberry sawfly, currant sawfly, and imported currant worm are all common names for *Nemastus ribesii*, an uninvited guest from Europe that eats the leaves of red currant plants and gooseberries. In June, we might pick off dozens of these pesky green larvae. Sawflies are species in the same order of insects (Hymenoptera) as bees and wasps, although they don't sting. In the spring, after the currants and gooseberries have leafed out, the overwintered pupae in the soil emerge as adults and mate, and the female lays her eggs on the underside of the leaf, along the leaf ribs. After one to two weeks, depending on the weather, they hatch and the larvae start eating the leaves. The adult tends to lay her eggs near the bottom of the plant and often centrally located, so in the spring that's a good place to look for leaf damage or even eggs. Depending on the geographic location and climate, there can be multiple generations per growing season. In northern Vermont, we have two generations.

Since we experienced an outbreak a few years ago, we generally scout around our red currant and gooseberry plantings every few days in June and July to look for and collect any larvae. John recommends that we remove the larvae to behind the barn, far away from the bushes, too far for them to crawl back. Knocking the larvae to the ground doesn't solve the problem, because they will march back up the plant and start feeding again. We've watched them do it in a matter of minutes. They're voracious eaters. One of the reasons we like to leave them alive rather than squish them is because a few of the larvae may already be parasitized. Certain types of parasitic wasps and flies will lay their eggs within the larvae. We want those parasites to emerge and join us in controlling the larvae. When the parasite eggs hatch, they eat the inside of the larvae. The wasp larvae then pupate and become adults that parasitize more sawflies—a good example of naturally occurring biological control.

A few years ago, we had a population explosion of the gooseberry sawfly. Something was out of balance. That was only the second time in about fifteen years that we've had such an outbreak. For years, we hadn't seen any evidence of them. And then, for whatever reason, their numbers skyrocketed. We made a couple mistakes that probably exacerbated the situation. When we first noticed them in the spring, we were complacent, thinking that nature would take care of itself. Which of course it will, but that doesn't mean it won't be detrimental for the crop or the plants in the short term. We were not diligent about

rethinking pests, invasive species, and other paradigms

FIGURE 10.4. These sawfly larvae, commonly called currant worms, can cause defoliation damage if found in high numbers.

picking off the first-generation larvae when we initially noticed them in early June. This complacency on our part allowed more larvae to pupate and become adults, which made the second generation that hatched in early July much worse.

The second problem was that before the currants started turning red, we netted them to keep the birds, especially the cedar waxwings, away from the red berries. Many birds will eat the larvae, and some birds, such as our barn swallows, prey on the adults; unfortunately, the net kept the larvae-eaters away. Also, the net prevented us from getting at the sawfly larvae once they started to reach high numbers. By the time we started to take the problem seriously, they were out of control. Before the late-season currants were ripe in later July and early August, almost every red currant bush and several gooseberry bushes had been completely defoliated. While we saved our crop from the waxwings, the defoliation took a toll on the bushes, and we had to prune out a considerable number of dead branches the next year, which meant fewer currants.

Besides ourselves and the birds, there are other predators of the larvae in the bushes. Turning over a leaf with tiny holes on it, we find not the sawfly larvae but a small alligator-like ladybird beetle larva crawling underneath. Maybe the beetle did the work for us. Predaceous stinkbugs also lurk in the leaves. With their piercing and sucking mouthparts, they're good predators of the currant worm. Predaceous ground beetles will eat the pupae in the soil. There are also fungi and diseases that will attack the larvae, the pupae, and the adults. We guess it's not that easy being a gooseberry sawfly.

A conventional farmer faced with the same problem would most likely reach for a chemical solution. One of the issues with using most insecticides is that they kill off not only the sawfly larvae but also the predators and parasites of the sawfly larvae. Predator and prey therefore never come back into balance, which means there is increased potential for an outbreak the following year, thus validating the need for chemicals every year—a good dependency-creating scheme by the chemical companies. When we experienced our outbreak, we were also only weeks away from the harvest, so pesticides would have persisted on the fruit, or *in* the fruit if we had used a systemic pesticide. Pesticides can also contaminate the soil and leach into runoff after a rain, disrupting more of the ecological balance as they kill or weaken soil and aquatic organisms, large and small. There is very little understanding of how pesticides interact with our own gut flora, which have been shown to play a huge role in our health.

Most people think that because a pesticide is approved for organic producers, it's safe to use. A woman recently emailed us about another type of sawfly larvae on her serviceberry bush that she sprayed with spinosad (a.k.a. Entrust). Entrust is approved for organic growers, and it's made by Dow Chemical (now DowDuPont), the same people who gave us Agent Orange and DDT. Hence our sarcastic refrain, "In Dow, we Entrust." Spinosad is also a broad-spectrum pesticide, which means that this woman was also killing off predators and parasites of the sawfly, as well other nontarget insects like bees, and likely messing up the local food web in other unknown ways.

Which brings us to the other reason we don't want to smoosh currant sawfly larvae. We're trying to learn to coexist with other organisms (including having compassion and respect for what are commonly termed pests and invasives) instead of mindlessly killing species that coinhabit the planet with us. We're questioning this knee-jerk reaction of killing insects and other things that vex us, questioning the dogma about eradicating invasive species and insect "pests," and trying to mindfully observe and value the plants and animals around us.

Respecting biodiversity goes beyond a theoretical philosophy or scientific discussion. We want to make it part of our day-to-day living. It's not always easy, though. We still swat at mosquitoes and deerflies when they land on us. But nature can be harsh, and we are nature. Sometimes we have to protect ourselves or prevent an outbreak. Instead of killing, however, we much prefer rescuing a beetle or bee from our water barrels when they occasionally fall in. It makes us feel good, and so does being pesticide-free.

Bishop's-Weed, the Bane of Gardeners

You can't fight bishop's-weed, also known as goutweed or ground elder, with old-fashioned weeding. When weeded, the green leafy aboveground part of the plant breaks off from its tenacious network of underground rhizomes, which quickly grow into new hardy plants.

We heard the only way to get rid of bishop's-weed, a member of the carrot family, was to eat it. It's a good spring and early summer veggie with lots of vitamins and minerals. It's also high in antioxidants and has several medicinal benefits. Unfortunately, we have large patches distributed in shady areas around the house and buildings that would easily withstand our eating it for control. It grows around our stone wall, the old flower bed, the back barn, and the back porch. Bishop's-weed was another plant brought over by the early colonists for food and medicine. It's also been sold as an ornamental around the world. Bad idea.

It grows well in shade and outcompetes everything, even grass. In the spring and early summer, it makes for a lush green ground cover. In the late summer, it often looks brown and ragged, but sends out new growth and looks green and lush again in September and early October, which helps it outcompete other plants. We've mulched it with cardboard and landscape fabric, but that's a temporary fix. Like quack grass, it comes back quickly, eventually moving through the fabric or finding gaps in the cardboard. And then it looks like you didn't do anything at all.

When Nancy was a kid, back around second and third grade, she and her friends made "vinegar" with gout weed. That's what they called it anyway. Kids would put the leaves in quart jars with water and set them in the sun all day. The next day. they'd drink it up. They should have called it an herbal sun tea and sold it instead of lemonade at their lemonade stands. It didn't taste too bad. Now, whenever she pulls up gout weed, the aroma reminds her of their drink. It has medicinal properties, and as the name implies, it's an old remedy for gout and arthritis. It may also act as a mild sedative!

When we finally realized we couldn't possibly pull it all up or eat it or sheet mulch it away, we tried a different approach. As with the reed canary grass, we've been trying to find other plants to compete with it. We planted hostas at the base of the stone wall that are now growing thick and large and outcompeting the bishop's-weed. On the stone wall itself, we've planted a pink bistort cultivar that has been holding its own and spreading along the wall. Its pink flower heads in the spring make an attractive bee and wasp plant. As for the rest of the bishop's-weed on the farm? Well, we're learning to live with it and hoping that this herbal remedy comes back into fashion soon. Then we'll be rich!

CHAPTER 11

the bees' needs

Pollinator conservation is near and dear to our hearts. It also makes good economic sense for fruit farmers. Pretty much everything we grow on the farm starts as a flower and needs a pollinator to bring pollen from the male flower parts to the female flower parts in order to produce fruit. Most of the native plants that are so crucial to the ecological functioning of our farm are also dependent on pollinators. While insect pollinators can include butterflies, moths, flies, beetles, wasps, and more, we recognize that native bees such as bumble bees, mason bees, and ground-nesting bees (even more so than our honey bees) are our most important farming partners. We want to uphold our end of the partnership by enhancing and increasing farm biodiversity with pollinator and beneficial insect habitat. That has been one of our focuses over the years. Providing food and nesting habitat for native bees and creating a pesticide-free environment has undoubtedly increased the viability of our farming business while also helping to heal and regenerate the ecosystems that we steward.

We have come to the conclusion through many years of observing the bees visiting our fruit flowers that the native bees are more important for our fruit pollination than honey bees. We enjoy honey bees and

FIGURE 11.1. A native bee gathers pollen from Jacob's ladder flowers.

their honey, but let's face it, they are basically non-native, domesticated bees originally from Europe. They might even be considered invasive if you consider their ability to go feral and the fact that they compete (and in some cases outcompete) with native species for the same resources. We appreciate the honey bees visiting our apples and pumpkins, but native bees such as plasterer or squash bees are doing the larger share of the work and not getting much credit. On other fruits such as cherries, plums, honeyberries, blueberries, and currants, we rarely see a honey bee. This might be due to floral preferences, early flowering, or poor weather when the honey bees are staying in their hives and enjoying stored honey. The spring-emergent native bees tend to fly in cooler, rainier, windier weather and are attracted to many flowers that the honey bees pass up.

Pollinators in Peril

Many people are aware that honey bee populations are stressed as a result of pesticide exposure, parasites and diseases, loss of habitat, and climate change. The stresses are real, but we don't worry too much about them, because they have beekeepers like ourselves to help them along. We can overcome winter losses and support our bees by splitting hives, buying new queens, treating for mites, and otherwise adapting our beekeeping practices. While honey bees' iconic stature draws society's attention to environmental causes, we joke that the slogan "Save the Honey Bees" is the insect equivalent of "Save the Chickens" if it came from bird conservationists.

A more pressing problem is the plight of native bees and other wild pollinators because for the most part, they're on their own. However, what is good for honey bees is also good for native pollinators, generally speaking, so the recent not-so-accurate media coverage that focuses on honey bees can be a good thing. But drastically increasing the honey bee population or the number of poorly managed hives in a given area can lead to competition with our native bees for floral resources as well as exacerbate the amount of parasites and diseases being shared. Everything is connected, and flowers visited by different bee species can become a channel for diseases and parasites, kind of like a dirty doorknob!

There are approximately 275 native bee species in Vermont and around 4,000 in the United States. Many of these species are thought to be in decline. The rusty patched bumble bee, for example, was put on the Federal Endangered Species List in 2018. Once common in twenty-eight states, it has suffered from a diminishing range according to records, and now this bumble bee is found in only thirteen states. Native bees and other wild pollinators are farmers' unsung partners in many cases, providing pollination services for farm and garden crops as well as the native trees, shrubs, and wildflowers of our natural landscape. Social bumble bees and solitary bees, such as the mason bees, digger bees, sweat bees, leaf-cutting bees, sunflower bees, and squash bees can be more efficient pollinators than honey bees and help keep our farms and gardens bountiful. They need our support now more than ever. We prefer a rallying cry of "Save the Native Bees!"

Pesticide-Free

There are several action areas we've focused on over the years, such as creating nesting and overwintering sites and ways to increase season-long floral resources, but one of the most important practices for us is being pesticide-free. Not all pesticides are equally toxic to pollinators, but most will have harmful effects on nontarget organisms and insects and can be very disruptive to farm and garden ecosystems.

One example is a harmful class of insecticides known as neonicotinoids (neonics), which have become widely used since the mid-1990s.

FIGURE 11.2. Pesticide-free sign posted below a cavity-nesting bee box with straw inserts.

These are also available to homeowners in lawn and garden products. Neonics are used on most conventionally grown corn and soybean farms as a standard seed treatment. They are systemic pesticides that once applied are expressed throughout the plant, including the nectar and pollen. Even low neonic concentrations in nectar and pollen have been shown to interfere with navigational systems of bees when ingested, as well as their autoimmune responses, reproductive potential, and other aspects of being a bee. These sublethal effects can cause additional stress to already stressed bee populations. It's not only neonics that can harm pollinators. Some organically approved pesticides are also highly toxic to bees!

Our training, education, and thinking over our lives have led us to acknowledge how little we humans really know about the short- and long-term effects of pesticide use on farm and garden ecosystems. Insecticides, fungicides, and herbicides are all biologically active compounds that can have negative effects on nontarget organisms such as beneficial insects, soil microbes, and important plant-fungi interactions. We humans also have little understanding of the effects of pesticides on another complex ecosystem: the human body, including our gut biome! Our personal philosophy is not to use them on our farm and to consume as little as possible of them in our diet. We believe in the precautionary principle, which means precautionary measures are taken when scientific issues around cause and effect with regard to these toxins are not fully understood. It is a shame that our chemical companies and regulatory organizations are so quick to release things into the environment without adequate understanding or respect for the risk of long-term consequences, and that consumers are so quick to use them.

Bumble Bees and Their Nesting Needs

Bumble bees are an example of native social bees and are the pollinator workhorses on our tree fruits, currants, gooseberries, blueberries, raspberries, and greenhouse tomatoes when we grew them. When we used to grow more pumpkins and other cucurbits like cucumbers and watermelon, bumble bees and squash bees were the main visitors we observed on the flowers. Bumble bees can build up colonies over the summer that include as many as three hundred workers if conditions are good. In the late summer, the original queen lays eggs that will become new queens or drones. These new virgin queens emerge in the late summer and autumn and mate with the drones, ideally from another colony to avoid inbreeding. Only newly mated bumble bee queens overwinter. Old queens, workers, and drones, having all done their jobs, die out before winter.

Toward the end of April here in northern Vermont, about the time the shrub willows bloom, we see the first new bumble bee queens emerge from their protected overwintering places on the farm. They are out and about, looking for food and nesting habitat. The fat reserves that got them through winter hibernation start dwindling, and they need nectar for energy. To find suitable nesting habitat, they cruise low near rock walls and potential cavities in the ground or in the old hay bales we intentionally leave lying around. Abandoned mouse nests are often a favorite place to establish a colony, so we try to create artificial mouse nests in wooden boxes or by putting hay bales on pallets with rain covers. Any old mouse nest material found around the farm is put in these nest spaces because bumble bee queens have been shown to be attracted to the smell of mouse urine!

If you happen to see a queen bee with pollen on her legs in the early spring, it means she has already chosen a nest site and is beginning to provision it with pollen for her larvae. In the spring, this queen does all the work. She collects pollen and nectar, makes the waxy brood and honey pots, and incubates the eggs until the first generation of workers emerge in less than a month. Talk about a supermom! After the workers start helping with foraging and brood-raising chores, the queen can stay at home and focus on laying eggs for the colony.

FIGURE 11.3. Tricolored bumble bee visiting a phacelia cover crop.

FIGURE 11.4. Bumblekultur is planted with a diversity of summer flowers. *Photograph courtesy of Michele Lamazow.*

On the farm, we generate a considerable amount of prunings from our fruit trees and berry bushes. While many people burn theirs, we hate the idea of putting all that carbon into the atmosphere when we can put it to use on the farm to feed our soil. We make long tall mounds of the prunings and woody scraps and cover them with soil or composted manure to turn them into hügelkultur (see chapter 5). We call them "bumblekultur" on the farm because we're trying to create possible nesting and overwintering habitat for bees. Over time, the prunings break down, leaving a smaller mound high in organic matter, but for the first several years, there are a lot of cracks and crevices that can serve as bee habitat. We grow squash, gourds, food crops, or cover crops like buckwheat, nasturtiums, and phacelia in the first year (while the pile settles). In later years, we plant perennial and annual flowers. Not only does the three-dimensional aspect help for native bee nesting and overwintering sites, but planting on the mound provides additional floral resources besides. And we're sequestering carbon. Now, that's stacking functions!

Solitary Bees and Their Nesting Needs

Solitary nesting bees make up the majority of our native bee species. They live for one season and work alone, laying their eggs in chambers in the ground (ground nesting) or in small tunnels in trees, cracks and crevices in buildings, or in plant stems (cavity nesting). Wherever the eggs are laid, the larvae are provisioned with pollen balls to feed on. Bees are vegetarians and are adapted to collect pollen, as opposed to their more ill-tempered cousins, wasps, which are carnivores with a penchant for nectar as an energy source. Solitary bees are generally out and about as adults for only a few weeks during the season. In that short time, the females mate with the single-minded (some might say shiftless) males and begin to create the next generation. The females work alone to fill hollow stems and tunnels or underground chambers with pollen balls and eggs. The longest portion of solitary bees' lives is spent as larvae wallowing in a pollen ball of food in a cozy nest. Depending on the species, solitary bees overwinter as last-stage larvae, pupae, or adults. Different species emerge at different times over the growing season and have specific plant and nesting preferences. For example, blue orchard bees emerge in early spring and are especially effective pollinators for apples and other fruit trees. They lay their eggs in preexisting cavities like beetle bore holes in tree trunks or hollow stems. Squash bees (pollinators of cucumbers, squash, and pumpkins) are a good example of ground-nesting bees that emerge as adults in the summer in time to get to work. You can often see them in the early morning. The males can be seen hanging out in the flowers, waiting for the females to come by, reminiscent of guys who go early to Ladies' Night at the bar.

Putting out bee boxes for cavity nesters is a great way to help establish and grow populations of different bees, such as blue orchard bees or leaf-cutting bees. We prefer to use disposable paper straws, six-to-eight-inch (15 to 20 cm) cardboard tubes, or hollow plant stems in our cavity-nesting bee boxes. That way we can clean them and replace them every year to avoid the possible buildup of parasites and disease. Drilled blocks without replaceable straws can be a problem and are more difficult to maintain. The straws, tubes, or plant stems should be placed within wooden boxes or other handy holding devices (for example, cut-open plastic water bottles) that will keep them dry.

These boxes are generally attached to the south side of posts, buildings, or trees. If the tube is accepted, a female mason bee will put a pollen ball in the end of the tube and lay her egg on it. With a little clay mud, she will seal up the egg and pollen ball in a small chamber. She will continue to place pollen balls and eggs in the tube, sealing off each chamber as she does. When she's done, there will be six or so sealed chambers with pollen balls and eggs in each straw. Leaf-cutting bees line their nest holes with small cut pieces of leaves. We are delighted when we see the half-moon holes cut out of our rose leaves. These are your leaf cutters in action. Please don't reach for the pesticides if you see these holes in your rose leaves!

Different species of bees prefer different-size holes. Blue orchard bees are attracted to holes around 5/16 inch (8 mm) in diameter. To attract a variety of species, use a variety of hole sizes. Cardboard straws are available from suppliers, but we also use hollow stems (phragmites) and sticks with pithy centers (sumac) that we drill out with the appropriate drill bit. Generally, the tubes should be replaced annually to prevent the spread of pests and diseases.

Mason wasps or grass-carrying wasps may also use the holes designed for the bees. They bring in tree crickets (in the case of grass-carrying wasps), or caterpillars (in the case of mason wasps), lay

FIGURE 11.5. A bee box built for cavity-nesting bees and wasps. This one has mason bees, mason wasps, grass-carrying wasps, and leaf-cutting bees all living in the same apartment building. Wasps are great predators of many garden pests.

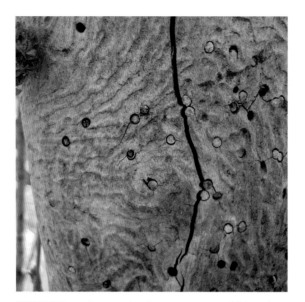

FIGURE 11.6. Cavity-nesting bees are using old beetle holes in a dead apple trunk for nesting. The capped holes are occupied with growing bee larvae.

FIGURE 11.7. Small carpenter bees make holes in old cut raspberry canes.

their eggs on them for their carnivorous larvae, then seal the hole with mud or grass. Although they're not great pollinators, they're beneficial in that they use many of our garden pests as food for their larvae.

Leaving the plant stems and not mowing areas with elderberry, sumac, joe-pye weed, and other hollow or pithy stems is another way to provide nesting habitat for cavity-nesting bees. The small carpenter bee (*Ceratina*) is a quarter-inch-long (6 mm) bee that loves to dig tunnels in the pith of last year's raspberry canes. We prune six inches above the ground level to leave vertical stems aboveground. We also leave our prunings in piles that are accessible to these little pollinators. Celebrate the scruffy with *Ceratina*!

Many ground-nesting bees, including squash bees and *Colletes* (cellophane or plasterer bees), prefer open sandy soil to tunnel and build their nests for egg laying. They especially appreciate a south-facing slope. Leaving these open spaces in fields, gardens, or backyards is a great way to encourage these delightful and docile bees. They can aggregate in favorable spots, with hundreds of holes in a small area. For years, we have seen only a few plasterer or cellophane bees around in the spring on our farm. To encourage a larger population, we created sandbox patches this past year and introduced bees caught at other sites.

Occasionally, a news article about using little flying drones to solve our pollination problems comes across our news feed. This is a ridiculous reflection of technological fundamentalism, whereby it is believed that we can invent ourselves out of our crises without basic behavioral changes. Of course, someone is working to make a profit on the pollinator dilemma through technology. We would rather work with the bees we already have. We are committed to figuring out how to be better native beekeepers.

Enhancing Floral Resources for Pollinators

When native bees emerge from overwintering, they need to feed and gather pollen for their soon-to-be-laid young. Some of our earliest-blooming shrubs are willows. These include purple osier willows (a favorite of basket-makers), pussy willows, black willows, and dozens of others. One of the first things we did after we bought the farm was to allow the willows, birch, box elder, and other trees to grow up along the seasonal stream that runs through the back pasture. We are glad we did, because this gives us an abundance of early-season nectar and pollen.

Spring bulbs can be another important source of pollen and nectar during these early spring days. Crocuses and daffodils close to the house often bloom earlier than those planted out in the yard. We enjoy seeing the native bees taking advantage of them. We've also planted Siberian squill in the apple orchard on the side of the house. They've become a favorite early-spring meal, especially for the blue orchard bee. Providing nectar and pollen resources during this early-spring "shoulder" period can help not only the native bees but honey bees as well. Every early bee is looking for energy in the nectar and the protein found in pollen to feed their young.

The blossoms of early fruit, such as honeyberry and serviceberry, are the next to open on the farm. Serviceberries are native to Vermont and are common around the edges of woodlands. From mid-May through early June is when most of our fruit trees and berry bushes are in bloom. Many native trees and shrubs are also blooming. Many flowers such as blueberries and tomatoes require buzz pollination furnished by bumble bees. This is when the bee vibrates the flower at a certain frequency to encourage the pollen to release. Also, the shape of the blueberry flower has evolved to suit such native pollinators as bumble bees with their longer tongues. That's why enhancing the bumble bee population around your blueberry patch can help your blueberry production. Because bumble bees require season-long floral

TABLE 11.1. Beginning blooming periods for pollinator floral resources.

Beginning Bloom Periods*	Plants
Late April / Early May	Willow shrubs (different varieties), speckled alder, maples, Siberian squill, daffodils (near house foundation).
Early to Mid-May	Plums, cherries, Cornelian cherry dogwood, honeyberry, currants, gooseberries, saskatoons, chokecherry, lungwort.
Mid to Late May	Apples, pears, blueberries, clove currants, Jacob's ladder, narcissus, ajuga, dandelions, creeping Charlie, creeping phlox, bistort superba cultivar, dyer's alkanet.
Late May / Early June	Aronia, lilacs, bleeding heart, invasive honeysuckle, Labrador tea, lily of the valley, buttercup, iris, peony.
Early to Mid-June	Dogwoods (red osier, yellow-twig, and pagoda), highbush cranberry, nannyberry, chives, lupine, black cherry, black locust, wild raspberries and blackberries, wild phlox, comfrey, ninebark, baptisia, clovers.
Late June / Early July	Virginia rose, clovers, primrose, catnip, mountain ash, hawthorn, common milkweed, swamp milkweed, lady's mantle, forget-me-nots, centaurea, basswood, sumac.
Early to Mid-July	Greenhouse raspberries, silky dogwood, yellow loosestrife, meadow rue, blue vervain, bee balm, butterfly weed, wetland rose, Saint-John's-wort, Queen Anne's lace, echinacea.
End of July / Early August	Buttonbush, globe thistle, early goldenrod, joe-pye weed, boneset, peppermint, flat-topped aster.
Mid-August to Early September	Late goldenrod species, sneezeweed, purple asters, New York ironweed.
Late September to October	Maximilian sunflowers, sunchoke, witch hazel, late aster.

* Most plants have a peak bloom for a few weeks. Others, such as creeping Charlie, can bloom all summer.

resources, it's important to have food available during July, August, and September, after the blueberry flowering period is over, if you want bumble bees around the following May and June.

We haven't mowed much of our pollinator sanctuary meadows for almost ten years. Thus far, we haven't had a problem with successional trees and shrubs moving in. If and when we do, we will time our mowings for the end of October, after the monarchs have pupated and begun their migration back to Mexico. October is also after the late-season goldenrod and asters are done providing nectar and pollen for pollinators. By then, they have become seed heads that produce food for migrating and local birds. The stems of many of these plants are also great nesting habitat for cavity nesters.

We've planted most of the native trees, shrubs, and perennial flowers in the pollinator sanctuary to support our native pollinators. These native plants also support native insects that eat the plants, which in turn support birds and other wildlife. The guild plants provide additional services to the fruit trees. We think a lot about the bloom time when we plant. Locust and basswood trees, for example, bloom from the beginning of June (locust) to the end of June or early July (basswood). They provide additional pollen and nectar for the bees without competing with our fruit trees for their attention.

Although we've been slowing down on the number of fruit trees and berry bushes we're planting of late, we continue to plant shrubs and perennial flowers favored by bees. We're still adding more guild plants to our polyculture orchard, but also adding to windbreak hedgerows and wildlife corridors. It's important to note that bees prefer white, pink, yellow, and blue flowers. We tend to focus on those colors when planting perennials and annuals, but we have a few red flowers such as red bee balm, which attracts hummingbirds and butterflies. Because many bees tend to practice floral constancy and stick with one type of flower while foraging, it's better to plant larger patches of single types of flowers rather than one here or there.

Several years ago, we conducted a study funded by SARE to look at using cover-cropping as a way to enhance floral resources. A short summary video about this research can be found at www.thefarmbetween.com/resources. Although we have become mostly a no-till perennial polyculture farm, as former vegetable farmers, we know the importance of cover-cropping for enhancing the soil, preventing erosion, competing with weeds, and other benefits. Our study looked at adding another important cover crop function, that of nectar and pollen resources for pollinators. We conducted replicated trials of buckwheat, phacelia, and a perennial conservation mix. We found that cover-cropping with species like buckwheat, vetch, and clover and letting them go to flower is a great way to enhance these bee resources while at the same time doing your soil a favor.

Our perennial conservation mix did not fare well in the short term, and it took a couple of years to get established, but in the long term, it has turned into a wonderful bee resource—especially the early-season lupine and the late-season perennial Maximilian sunflowers.

FIGURE 11.8. Growing an annual bee-forage-mix cover crop provides nectar and pollen for bees and other pollinators.

While our annual plots from that research have long since been converted to perennial plantings, our perennial conservation mix plots have remained as permanent bee resources.

Honey Bee Keeping

Recently, there's been renewed societal interest in keeping bees, perhaps because they are in the media so often with stories of colony collapse disorder or other headline-grabbing news. Many people enjoy honey and other bee products. A few are also interested in apitherapy, using bee venom to treat various chronic ailments. Some may also want to keep bees to enhance pollination in their farms and gardens and to learn more about social insects.

Honey bees are fascinating organisms, and beekeeping opens the door for lifelong learning about them. They were originally introduced to North America in 1600 by early European colonists. Learning about them, helping to take care of them, and gathering their honey and wax can be fun and profitable. We have been keeping bees for many years. John especially enjoys spending time visiting the hives, watching their behavior, catching a summer swarm, and selling and using the honey they make. It's really not about the honey, though. It's about the bees, and to make beekeeping work, especially for backyard enthusiasts, you need to be interested in the bees and in caring for them.

There are a variety of reasons why honey bees are in trouble these days, including many of the same threats affecting native bees and other insects. Pesticides top our list. Neonics, fungicides, and other chemicals are used on the plants bees visit. Most beekeepers also use pesticides in the hive for treatment of mites and other pests and diseases. Although pesticides may not kill the honey bees outright, they can stress them and make them more susceptible to pathogens (foulbrood, viruses) and pests (*Varroa* mites, tracheal mites).

Honey bees can also suffer from poor nutrition. Many beekeepers take extra honey and pollen from the hive and feed the bees sugar syrup and pollen patties in the fall and spring. Let's be clear: sugar syrup is not nectar, and no one really understands what it does to the bee's gut bacteria and overall health. We used it the first year when we didn't think our new hives had made enough honey to get through the winter. We didn't take off any honey either. Since then, our bees have always made enough honey to overwinter and we take only the excess.

Unlike native bees, honey bees overwinter as a colony. They keep warm by buzzing in a tight cluster and use the honey they've stored for food energy. In late winter, the queen starts laying more eggs and the bees need honey and pollen to feed the developing brood. That's why leaving them with enough nutritious real honey is critical for overwintering bees. In sparse years, we take less honey for ourselves, to make sure the honey bees have enough. We also store a few frozen frames in our freezer just in case a hive is running low in the late winter and needs extra food.

The evolution of industrial agriculture has also spawned industrial beekeeping. Bees are trucked around the country in order to pollinate large monoculture crops. Almonds in the Central Valley of California are a good example. There are almost one million acres of almond orchards in California, where the recommended stocking rate for pollination services is two hives per acre. During a two-week bloom time in February, billions of bees are brought together from all over the country to pollinate these almond orchards. For two weeks, the honey bees live on this monoculture blossom supply. Before and after arriving in sunny California, the bees are often fed sugar syrup, further stressing already stressed bees. In addition, fungicides and other pesticides (depending on the crop) may be sprayed while the bees are foraging, so it's no wonder migratory commercial beekeepers lose a good percentage of their hives every year.

The genetics of the honey bees also matter. Many varieties are more susceptible to mites and the cold winters of Vermont. We get our bees and queens from local northern Vermont beekeepers. This ensures winter hardiness. We often split hives either intentionally or unintentionally (when they swarm), which can break the bee brood cycle as well as the mite brood cycle.

As early as 1622, feral honey bee populations were noted in North America, and these populations expanded geographically with the expansion of the colonists. These feral colonies flourished until the accidental introduction of *Varroa* mites in the 1980s, which decimated wild populations. Domesticated hives were also severely affected, but beekeepers moved quickly toward using a chemical treatment approach that is standard today. There is evidence that feral populations may be recovering, which means genetically either they are evolving to "live" with the mites, or the mites are adapting to not kill off their host, or both are adapting to each other. Bee researchers and some beekeepers are also trying to develop varieties with genetic traits that are better adapted to *Varroa* mites. For example, there are hygienic bee varieties that search out and remove parasitized larvae, thus reducing mite populations in the hive.

In 2006, several large migratory commercial beekeepers reported the sudden death of many of their hives. Most of the bees from a given hive had vanished with the occasional exception of the queen and a few of her attendants. The term "colony collapse disorder" (CCD) was used to describe this condition. CCD shouldn't be confused with death of the overwintering hive, which can happen frequently in Vermont. Winter losses can be high, and the causes are numerous, including *Varroa*, insufficient honey, pesticides, or the combination of all of these. CCD is also believed to be the result of a combination of various bee stressors.

New beekeepers have lots to learn. Patience and keen observation are important parts of the process. Getting hive supplies, equipment, and the bees themselves can also cost quite a bit up front. Besides the up-front costs, time is an important component—time to get into the hive and time to read and learn more about them. You need a location that is easily accessible and safe. In our area, bear-proofing is a must.

It's important that the resources and ecology of an area are able to sustain colonies, that the addition of thirty thousand bees will not outcompete the native bees. Conducting a survey of the plants in the area is a good start. We do a weekly walk, noting the different species, their flowering time, relative abundance, and which pollinators are visiting. A survey is a great exercise for increasing observation skills and creating a greater awareness of the natural world. Planting additional trees, shrubs, and flowers—especially ones that bloom during times when not much else is in bloom—should be considered. Leaving areas with wildflowers unmowed is also a good management strategy.

Keeping honey bees also requires physical strength to lift deep supers and honey supers. And finally, you can't mind too much about getting stung. Bees can sense if the beekeeper is afraid or angry, and that can get them riled up too. Beekeepers need to be calm and collected to get the most enjoyment from beekeeping. Part of the challenge is getting into the right frame of mind.

Commercial Bumble Bee Keeping

Raising and selling bumble bee colonies has recently become more popular as a commercial business operation, especially with the advent of growing tomatoes and other crops in high tunnels. Commercial bumble bee colonies are raised in special facilities and shipped by mail to the farmers. Many organic raspberry farmers now use a fine-mesh fabric screen on their hoophouse sides to keep out spotted wing drosophila. Since this makes it impossible for bees to get inside the hoophouse, they also purchase and bring in bumble bee colonies for pollination. A major concern is that these bumble bees have been shown to harbor parasites, viruses, and other diseases that can spread to honey bees and other native bees. In open-air situations, we are not in favor of people bringing these colonies onto their farms. When a greenhouse or hoophouse is sealed by screening, however, there is probably less chance that viruses and other diseases will be spread by these lab-raised nonlocal bumble bees.

We have been experimenting with raising bumble bee colonies from locally wild-caught queens through a USDA Specialty Crop

FIGURE 11.9. The inner workings of a bumble bee colony shows a large queen and several workers. A honey bee (now dead) was trying to rob the nectar.

Block Grant. Locally raised bumble bees might alleviate some of the disease problems mentioned above. This past spring, we raised twelve queens in individual containers in a dark heated closet and fed them sugar water and honey bee pollen. Three of the twelve started colonies, and we released the ones that didn't after a couple of weeks. We put the colonies outside once the first generation of workers hatched. We watched them throughout the season. They didn't really grow into large colonies, but it was fascinating to see through a Plexiglas cover their development throughout the growing season and share that with our visitors.

While we enjoyed raising bumble bees, we think creating habitat on the farm where native bees will thrive is a much better idea. Eliminating pesticides and creating a biodiverse landscape with season-long floral resources and nesting habitat for native bees will bring immediate results.

CHAPTER 12

sharing the farm and farm products

We realized early on that we didn't want the farm to be only for us. It was meant for sharing. We hosted several family reunions from both sides of the family here, for example. With all our animals and fresh produce as well as the local beauty, it was a great vacation and reunion spot for our siblings' families, especially all the kids. Longtime friends with their families also made the trek to visit and check out the farm. As, of course, did the local community.

Over the years we've connected with nearby schools and summer camps that want to use the farm as an outdoor classroom for field trips and class projects, which has meant thousands of kids visiting the farm. One of our most memorable groups of campers involved five visually impaired children and their camp counselors. Preparing for their visit, we thought about ways to break out of our visually dominant presentation. We typically rely so much on sight perception that we often forget about our other senses. We came up with a variety of nonvisual ways the kids could experience the farm. We brought them to different flowers and herb patches to smell. We gave them green beans and blueberries to taste. We walked through the riparian zone and listened to the moving water and bees and birds. We had them weed by touch, which actually worked well around perennial shrubs. The campers were extremely curious and engaged.

At the end of the tour, we brought out our horse, Nellie. She was such a gentle horse. She wouldn't get excited or move when groups of children combed and brushed and touched her. It's always been a treat for us to watch the kids enjoying her, but this group of blind children was even more special. Their sense of pleasure and connection with Nellie moved us deeply. It reminded us to be open to the moment and enjoy the farm with all our senses and childlike wonder. We always learn from our visitors too.

We've also hosted a variety of Northeast Organic Farming Association (NOFA) and Rural Vermont workshops and tours on the farm, with topics ranging from keeping a family cow to grazing rabbits, pesticides issues, and pollinator conservation. We hosted a biodiversity-training workshop for organic inspectors here a few years ago. We regularly give presentations about our farm practices at the NOFA winter conference and other conferences. Our relationship with NOFA-Vermont has been very important to us. John was on the NOFA-Vermont board of directors and served as co-president. He's served on other Vermont boards, including the Vermont Vegetable and Berry Growers Association and the Intervale Center. He's always been a networker and a connector, reaching out at farmer conferences and meetings.

FIGURE 12.1. Students from the Big Picture Project at South Burlington High School learn about pollinators on the farm. *Photograph courtesy of Kevin Downey.*

In recent years, most of the student programs we've worked with have been at the college and advanced high school levels. Students in courses on ecological agriculture, fruit growing, composting, or permaculture have visited and worked on projects with us. We've also been involved in summer farmer training programs. One of the ways we offset the time and effort for giving farm tours is a work exchange. It's good hands-on experiential learning for our visitors as well and often takes the form of mob weeding, pulling up heavy landscape fabric, picking berries, planting perennial flowers, collecting milkweed pods, or other farm jobs. Having a couple dozen or more hands helping out while engaged in lively conversation turns a daunting job into something fun. We chalk this up as a win-win situation.

While we usually start these field trips by sharing our love of nature and the farm, we've also tried other techniques of late. For example, we often stop at a particular location to let the students observe and make notes about what they notice and what they might have questions about, which stimulates their sense of curiosity and wonder rather than passively receiving the top-down teaching of a "talking head." Permaculture, ecology, and mindfulness traditions all emphasize the importance of awareness and observation, yet all too often our educational system spoon-feeds students. In our rush to condense the farm into a one-hour tour, we've also been guilty of this. It's efficient and people might gain a lot of information that way, but it's not necessarily helping people to observe and learn on their own. As teachers, we want to help others learn. We try to remember that when we share the farm.

Agritourism Adventures

Around 1995, we participated in an agritourism workshop that gave us plenty of ideas for increasing business for our farm in the form of tourism. Despite its location on our town's main road, our farm stand was doing a mediocre business at the time. We thought things like on-farm tours, farmer-for-a-day programs for kids, and a "must see" petting pen would help increase business. We received another SARE grant to support and evaluate our efforts.

Around that same time, we got a call from a woman who asked us to take in an animal she'd rescued from her next-door neighbors, a Vietnamese potbellied pig named Petunia. These neighbors of hers put Petunia in a halter that was way too tight and chained her outside. They'd been neglecting and mistreating her. When they spray-painted Petunia orange and painted her toenails pink for Halloween, the caller knew she had to do something. She didn't think; she just paid them fifty dollars to buy their pig. Now she needed to find another person to take Petunia. She wondered if we would give Petunia a home and reimburse her the fifty bucks.

Potbellied pigs were a fad in the 1990s. Touted as good house pets, they were adopted by people who treated them as if they were dogs, including training them and keeping them in the house. They supposedly stayed small and cute and could be house-trained. The demand drove a breeding business, but like many pet-breeding ventures, things got sloppy. Petunia grew beyond her potbellied genetics, which is why her halter no longer fit and why her owners kept her outside. She must have had a fair amount of full-size domestic pig in her genes. Domestic pigs are bred to grow fast and get fat. Petunia wasn't that cute anymore.

John agreed to take the pig for free. He felt bad for the desperate do-gooder but not bad enough to shell out fifty dollars for a pig we didn't really want. He made Petunia a pen and peeled off the halter, which had dug into her flesh in places. She seemed relieved.

FIGURE 12.2. Our brochure from the mid-1990s promoted our agribusiness ventures at The Farm Between.

But later, we thought, why not make Petunia the star attraction of our new petting pen, or pens in our case? We also had Freckles, a once bottle-fed lamb who grew up and was having lambs of her own; an angora goat; Muscovy ducks; laying hens; and geese that wandered the barnyard. We designed brochures and printed one thousand copies. We added a petting pen sign to our farm stand sign on the main road. We were ready to go.

The petting pens were part of our farmer-for-the-day program, which was great fun. Our teacher friend, who also had kids of her own, helped design and run it. She knew what activities to do with both preschoolers and grade schoolers, who as part of the program fed and met the animals, collected eggs, took a bug safari, and were part of the magic of the farm, complete with visits to an enchanted throne (an ornate high-backed velvet-seat chair) that sat on top of the hill out back. Fifteen years later, that hill became Knoll Orchard.

As we predicted, Petunia became the "star." She was about a year old during that first summer and weighed close to 150 pounds. In that regard, she was much smaller than a pig bred for meat production, but she was also much bigger than a true potbellied pig. She was black, not the usual spotted white or rusty red of domestic pigs. She wasn't cute, but she had a lot of personality. In other words, she was pretty mean and would try to nip your fingers if you reached in to touch her. Not a good choice for a petting pen with little kids who liked to put their fingers through the fence. Instead of having people reach in to pet her, we hung a long-handled hard-plastic bristled brush on her pen with a sign telling visitors to SAVE YOUR FINGERS. USE THE BRUSH! Petunia did like to be scratched.

Even with our grant and all our new strategies (the signs, programs, and brochures), we still had a hard time getting the word out about our farm stand and programs. The mid-1990s were a time before the widespread adoption of the internet, including Facebook and other social media websites. We gave a half dozen modestly priced farm tours, mostly to people staying at the nearby Smugglers' Notch Resort. In general, tourists seemed to just want to be entertained, and we quickly lost our enthusiasm for having tourists on the farm. The farmer-for-the-day kid's program ran a few times that summer, mostly for locals and friends. The following summer, our friend opened her own preschool business, so we had to let the farmer-for-the day program drop. We kept the petting pen, though, for our regular farm stand customers who enjoyed the animals and especially Petunia.

Looking back on the agritourism experiment, we realize just how much energy and enthusiasm we had back then that we could quickly fire up projects and run with them. After trying it that first summer, however, we realized it was going to take more than just a few brochures to make it work. Interacting with the general public wasn't always fun either. When we evaluated our priorities at the time, agritourism wasn't high on the list. It seemed like it was more of a distraction, so it was easy to let it drop, lesson learned, and move on. Cleaning out a back closet a while ago, we found the original box of brochures. There were probably well over eight hundred left.

FIGURE 12.3. Signage near our driveway shows The Farm Between logo. In the latter part of the summer that year, the nursery and gallery were open on Sundays only.

What's in a Name?

There's a lot of talk about branding these days and how important it is for the success of a business. How do you choose a name that promotes what you do? How do you keep it fresh and recognizable? The main problem with our name, The Farm Between, is that it doesn't convey precisely what we do. But then again, since we've remade ourselves many times over the years, we've never felt constrained to keep up with a product just because it was part of the farm name. We were a livestock-oriented farm when we started, so we're glad we didn't call it Hayden Meat Locker or something binding like that. We've seen this happen to new farmers who emphasize the meat they grow, chicken or pork, in their name only to get out of the meat business later in their farming careers.

In our case, as we remade ourselves by moving out of meats and vegetables into vegetables and fruits, and then fruits and fruit products, and now into wholesale fruit and a fruit nursery, we have only had to change our road sign and farmers market signs with a new tag line. Unfortunately, when people are cruising farmers markets or zooming past the farm at fifty-five miles per hour, they don't necessarily notice the tag line. We've also modified our logo to include apple trees and a bee on the cupola instead of the horse of our earlier logo. We still like our farm name. It's had the staying power that other names on our original list probably wouldn't have had.

Fruit Syrups and Snow Cones

Contact with the general public through our direct retail outlets—our farm stand, CSA, nursery, and farmers markets—has been an important part of connecting, sharing, and educating the public about our farm and farming practices. It gives us hope when people are curious about regenerative and agroecological practices and committed to supporting these practices by supporting us and, at the same time, feeding their families healthy food.

In ecology, diversity equals stability, and we believe this also holds true for marketing. A variety of products and a variety of outlets allow for flexibility and resilience in the face of change. We've always had a diversity of farm products, mostly due to our diversity of interests and the desire to feed our family homegrown pesticide-free food. If we were already going to grow vegetables for ourselves, why not grow a bit more and sell it to cover costs and workers? That was a good idea, but it could get a little crazy at times, keeping it all straight. We've grown different kinds of vegetables, fruits, meats, and produced a variety of value-added products such as syrups, fruit ciders, vinegars, and jams. We've even made ready-to-eat products: dill pickles, pocket pies, fruit between waffles, organic fruit snow cones, and fruit ciders (hot and cold). Some of these farmers market fair foods were more successful than others. For example, snow cones and fruit ciders were a huge hit. Having too many products, however, can often be problematic in that it complicates market preparation and setup.

FIGURE 12.4. John pours a black currant snow cone at the market.

About ten years ago, while we were sitting around a new farmers market in the village of Johnson with no one buying our radishes, lettuce, or rhubarb, we noticed the tamale vendor next to us had long lines; this started us thinking about value-added fair food. It seemed as though people were coming to the market for entertainment and to socialize. The next week we brought rhubarb juice in addition to rhubarb. When that didn't work, we came up with strawberry-rhubarb syrup that could be poured onto shaved ice. John had read a *New York Times* article about Hawaiian shaved ice, snowballs in New Orleans, and snow cones in New York City. We thought Vermont should have its own organic tradition. We found a hand-cranked ice shaver online, and a couple of weeks later, we had a line at the market, although not quite so long as the tamale line.

Our method for making syrup consisted of steam-juicing our fruit. We bought a stainless-steel steam juicer, which includes a bottom portion for boiling water. Stacked on that is a collection vessel with a colander-type vessel above that to hold the fruit. A cover tops the stack. We typically freeze our fruit first, which breaks the plant cells and allows better juice extraction. About five pounds (2.3 kg) of fruit and an hour (or a little less) of steaming time are used to process the juice. We then put the remaining fruit pulp into a jelly bag and press it with the cheese press from our family cow days to get more juice. This is especially important with fruit like black currants and elderberry, which still have a lot of good juice left in their skins even after steaming for an hour. We combine all the juice with organic cane sugar and heat the mixture to between 170° and 180°F (77° to 82°C). We sterilize the jars or bottles in the oven at 200°F (93°C) for 20 minutes and boil the caps. We add the hot syrup to the hot jars and seal. The low pH, high sugar content, and sterilization ensure a safe and relatively shelf-stable product.

Because our syrup contains no chemical preservatives, its color and flavor can change over time, which affects its shelf life. Strawberry-rhubarb syrup, for example, can lose color and flavor after several months. Black currant can last up to a year. For our elderberry, ginger, and honey syrup, we add organic lemon juice to bring the pH down below 4. Elderberries are not as acidic as rhubarb or black currants.

The summer after we introduced snow cones at the Johnson Farmers Market, we signed up for a couple more nearby summer markets. We also started selling bottles of our syrup. Snow cones were a big hit for a couple years in the smaller markets until we had to compete with new ice cream vendors; then our sales dropped. We decided then to focus on a single large market, the Burlington Farmers Market, where our diversified fresh fruits and fruit products had a wider audience.

John likes to come up with names for our new enterprises. When we first got into the fruit syrup snow cone business, he wanted to call it the Vermont Snowfruit Company as part of The Farm Between. He even went so far as to hire a marketing guy to come up with a logo.

FIGURE 12.5. Our selection of currants and gooseberries displayed at the Burlington Farmers Market.

Fortunately, we didn't end up liking the logo. That and further reflection about the confusion a new name would produce brought us back to sticking with The Farm Between as our brand.

John was on a roll, though, and came up with cartoon characters to help market our ready-made fruit products, including Cranky the Snow Cone, Sassy the Soda, and Professor Pickle. Professor Pickle was modeled after Nancy because she was a professor, and we were selling her homemade crispy dill pickles. For a while, we even toyed with the idea of making and selling T-shirts, hats, books, and other "Cranky and Friends" memorabilia. We could have gone big-time with Cranky and our hand-cranked shaved ice topped with organic fruit syrups. But did we want to? John was already starting to cringe at the moniker "Snow Cone Guy" that people (and kids) called him at the markets. He actually wants to be called "that practicing regenerative agroecologist." Would the name Cranky the Snow Cone Guy be better? After a year or two, Cranky, like the Snowfruit Company, faded back into the ideascape of John's mind.

A Word about "Biggering"

That brings us to the idea of "biggering." Several times we thought about expanding our syrup business. Sales were brisk and other retail businesses were asking to stock them. We had been keeping it small and direct retail so we could use our home kitchen. If we wanted to expand or sell wholesale to bars and restaurants, we would need to use a certified commercial kitchen. We went so far as to write a grant for creating our own commercial kitchen in one of the buildings we previously used for employee housing. We thought we wanted to take our syrup business to the next level. Specialty drinks and beverages were becoming popular, and our syrups made great cocktails, sodas, and dessert sauces. We figured we would make good money selling to restaurants.

We sent in the grant proposal but luckily didn't get it. After our egos cooled down from the rejection, we rethought the whole idea. Did we really want to spend more time in the kitchen? More time supervising employees? More time on sales and marketing of sugary syrups? Wasn't this a kind of sugar pushing? Yes, we used our delicious organic fruit, but then we added a sweetener like organic sugar or maple syrup, raw honey or our boiled-down apple cider. Soda made from our syrups contained way less sugar than a Coke, but was that really saying much? We let the whole biggering idea go and have been glad we didn't box ourselves into a production corner and away from our true passion of growing things.

Fruit Ciders

After a year or two of snow cone and syrup production, the apple harvest from our small Front Lawn Orchard had increased enough for us to make organic apple cider. Organic apple cider is still hard to come by, but when you think about all the sprays that are applied to conventional apples and how pressing juice concentrates those apples, it means conventional cider carries the high potential for pesticide residues in each glass. Once we started thinking that way, it's been tough to drink regular cider. We stick to our own.

We sold hot cider during the fall Burlington Farmers Market. On a cold day, hot organic apple cider with a shot of our elderberry ginger honey syrup was tough to beat. Pressing our own organic apple cider also prompted us to try making other products with it. We had wanted to come up with a no-sugar-added syrup, so we concentrated our apple cider to use as the sweetener. We cooked down a gallon of our organic cider (made with sweet apples) to one quart of concentrated cider. To this, we added juice from our steamed-juiced organic raspberries at about a 1:1 ratio. This became our no-sugar-added syrup for snow cones, and we bottled it for sale at the market.

We came up with another product using apple cider as well: "fruit ciders." These contained a blend of 20 percent steamed fruit juice from our organic fruit with 80 percent pressed organic cider. We hot-packed them into sterilized quart jars and sold them at the market. Our customers loved them. We realized early on that we needed more apples to meet the demand, so John started looking into local abandoned orchards as an option for accessing more apples to press. He found a couple, and soon we could count on an abundance of apples for pressing.

The Gnarly Fruit Collective was another of John's marketing ideas hatched around collecting apples from an abandoned unsprayed orchard in South Burlington. He found a few apple enthusiasts to help with the project. The idea was to sell apples to local juice makers and cideries and use the money to pay for their labor in rejuvenating the orchard. The Farm Between could also get a good crop for our needs by picking them ourselves and paying a small fee to the owners. By putting the orchard under The Farm Between umbrella, the orchard became certified organic.

We now had lots of apples to turn into cider and sold it mostly as fruit ciders. A sweet cider made with ripe apples combined with tart aronia, elderberry, raspberry, black currant, or beach plum juice created amazingly tasty drinks. And with no added sugar! We felt good

sharing the farm and farm products

FIGURE 12.6. Our fruit ciders sold at the farmers market.

about that. Each jar contained a lot of organic fruit, which made the fruit ciders a bit pricey. We thought of them as a nice treat on their own or mixed with seltzer to make a spritzer. What often surprised us was that people would spend a lot on our value-added products or snow cones, but balk at the price of the actual fruit. We're such a treat society. The average American consumes about sixty-six pounds (30 kg) of added sugar per year. Now, that's a lot of treats.

We sold the only local organic (and no-spray) fresh apples at the Burlington Farmers Market. We'd sort and shine the apples, but they were smaller than conventional and often had a few spots on them. We were surprised they didn't sell better. People at the market didn't seem to care about local organic apples or else they assumed that the other apple sellers were growing organic too. Or maybe they didn't want apples. One of the concerns farmers bring up now when talking about the Burlington Farmers Market is that it's become a big social event, a place to get lunch or pick up art cards. There are still dedicated produce buyers, but they are getting fewer and farther between. Like us, many of the farmers at the market have been diversifying their products, selling grilled sausage or breakfast sandwiches featuring their own meat, or popsicles and jams from their fruit to adapt to the changing times and shoppers.

Fruit Sovereignty: Our Fruit Nursery

After years of loading up the truck and taking the farm to the Burlington Farmers Market, we decided we wanted to bring people onto the farm again, those who shared our interest in fruit growing and gardening. Here on our wild farm, we could share our love of nature, including plants and insects. After all, we're farmers, not vendors.

We started our nursery for many reasons. It's estimated that about 35 percent of all households in the United States have a food garden, and that number seems to be growing, especially among young people. We really wanted to encourage others to grow their own, without pesticides and with nature in mind. Fruits contain so many good vitamins and antioxidants and are a great way to eat healthy. Yet organic local fruits are often hard to find unless you go out of your way. Growing your own without pesticides and increasing biodiversity on the home front is a great way to take care of personal health and environmental health at the same time. If we could encourage people to shrink their lawns, build habitat for wildlife, and help combat climate change, we would be providing a valuable service to society and the environment.

We also thought of a nursery as a means to increase our social and economic resilience. It would connect us to our communities and like-minded people, and increase farm cash flow in the spring, when we didn't have much cash coming in. Our nursery is open three days a week for about ten weeks in the spring as well as by appointment for people who can't make the regular hours. This has worked out well for our time management, and we can still be available as needed.

While our farm's entrance is on a main road, there's never been a visible spot along the road that would make sense for a farm stand or now the nursery. Customers need to drive in around back, which we believe has deterred customers from our farm businesses over the years. We have road signs, but most people like to see where they're going when they pull in. The addition of the apple orchard and shrubs and trees near the driveway make it even harder to see us from the road (see figure 1.3 on page 8). With the help of the website, most of our nursery customers are making this a destination place, rather than a spur-of-the-moment stop as they drive by. The busy 50 mph state highway doesn't encourage a quick-decision stop either, which was a hindrance for the

sharing the farm and farm products

FIGURE 12.7. Berry bush area at our nursery in early May.

early farm stand but seems less of a problem for the nursery. Also, the people who come to our nursery generally spend a lot more time and money than the customers at our vegetable farm stand years ago.

We can sometimes spend up to an hour or two with customers who want to learn about fruit growing and want to reinvent their own lawns and property. This kind of customer service has helped spread the word about our nursery so that we have been able to expand our customer base. We do provide on-site consulting as well, but we charge for that.

We love getting people to think beyond just adding fruit trees and berry bushes to their property by showing and talking about native conservation plants and pollinator plants. We want them to start thinking about the links between biodiversity and wildlife habitat and pest management and resilience. When we're talking with customers, we always try to inject a little ecology into the conversation. Some people are more enthusiastic than others, but still we're planting the seeds of ideas. We never know when or where they'll start growing.

One of the signs in our nursery area asks that important question HOW ARE YOU BRINGING NATURE HOME? inspired by Doug Tallamy's book *Bringing Nature Home*. When students visit, we use it as an educational prompt. We also hope to encourage other visitors and customers to think about their personal responsibilities in supporting their place in nature. And, hey, it might encourage people to buy more plants too!

Wholesale Markets for Fruit

Our focus on wholesale markets started with selling our black currants to a local winery, but until the past few years, wholesale accounts weren't a very large percentage of our gross sales. As the winery grew and other entrepreneurial beverage-type businesses started popping up in Vermont, we planted more currants and other fruit to serve that market. It's been exciting to see our farm and regenerative practices highlighted by the wineries, breweries, cideries, and soda makers that we sell to. We like that big once-a-season sale of the entire fruit crop to various local businesses.

Our rhubarb sales generally follow the same model, with several three hundred to four hundred pounds in sales per season to a food hub that divides the bundles into its members' shares. As we've moved into fruits such as aronia and elderberry, we've kept the wholesale model in the forefront. Our recent retirement from the Burlington Farmers Market in 2018 means we have transitioned to wholesale markets for the majority of our gross sales of fruit. We still sell retail fruit and products directly from the farm, mostly to stay connected with the public and to maintain our diversity.

The Vermont beverage industry continues to see increased growth. The period between 2014 and 2017 saw a 146 percent increase in sales, according to a recent Vermont Farm to Plate report written in behalf of the Vermont Sustainable Jobs Fund. This increase was driven primarily by Vermont craft beer sales. Many craft breweries are now looking to expand their roster of beer styles by adding fruit to their beer recipes. We have sold wholesale orders of black currants, aronia, beach plums, gooseberries, and elderberries to craft breweries. New cideries, wineries, and distilleries also offer opportunities for more fruit sales.

Staying Flexible

While the products we've grown and sold and the markets we've catered to have changed and evolved over the years on our small farm, we don't regret any of them. Changing things up has given us the chance to be lifelong learners, as well as given the farm its economic stability. Many of the changes have come about because of changing markets and our changing interests. The small-farm flexibility has allowed us to stay in business for more than twenty-six years in the face of evolving priorities, markets, and times. When Nancy was contemplating retirement from her off-farm job, we were lucky to get into the Vermont Farm Viability Enhancement Program, which provided business planning, marketing, and website services to give us a better handle on how the farm could be our sole livelihood.

We encourage all farmers, homesteaders, and home gardeners to stay flexible, resilient, and to persevere. We need more small farms and gardeners in this world, not fewer. That's why we share our practices, recipes, and ideas freely. It often takes only a little encouragement to send a person on a fulfilling pathway. This was one of the main reasons for us to start a nursery and write this book. It's so important for our health, our communities, and the planet to be growing nutritious food while regenerating the land. Even more important than scaling up our approach is the need to amplify it by sharing it with others with similar goals.

We've tried to keep pace with marketing changes over the years, but we're starting to feel our age, especially with regard to social media. In the early years, we used hard copy flyers and brochures and word of mouth to advertise our farm and events. We joined NOFA-VT, Rural Vermont, and other groups to network and be inspired. Later, community email lists sprang up, ranging from veggie and fruit growers' associations, to monarch enthusiasts, to our local community Front Porch Forum (an online community mailing list that keeps people abreast of local news), which has really helped us advertise locally in recent years. We've made websites and try to learn as we go. Website software has become more user-friendly these days, but we still hire a consultant occasionally to help with details and marketing changes (for example, getting the website mobile-phone ready). Facebook and e-newsletters are good, too, but they all take time and energy to maintain a presence in the public's eye.

Off-Farm Income: Supporting the Family Farm

Most farmers earn off-farm income, but for small farmers, off-farm income can make up much (and often the majority) of the household income. Data, graphs, and reports on this topic can be found at the USDA, Economic Research Service and National Agricultural Statistics Service. This has been common for a long time. People often assume that back in their grandparents' or great-grandparents' day, these past relatives were able to support a big family from the farm. When we delved a little deeper into our own ancestors' stories, we found they did other things to make money on the side.

Nancy's great-grandparents, for example, ran spring bark-peeling camps in Western Pennsylvania. Her great-grandmother and grandmother fed the crew that her great-grandfather organized. In the spring, when tree bark peels off easily, the men cut huge old-growth hemlock and oak trees. They then stripped the bark and used it in the leather tanneries because of the high amounts of tannins the bark contained. Train cars were filled to overflowing with bark, and in the early bark-peeling days, workers left the timber to rot. To put it bluntly, they cut down the forest and relied partly on that resource in combination with farming to make a living. This same great-grandfather was also a good carpenter and built houses on the side. The son, and even one of the daughters (Nancy's grandmother) were hired out as horse-driving teamsters to add to the family coffers.

A farmer grandfather on the other side of Nancy's family ran a still during Prohibition to earn extra money, and during the Depression, he worked on FDR's work crews while his wife ran the farm. John's grandfather in Ireland worked at Feahy's Forge as a blacksmith to pay some of the bills. Could they make a living on their small hardscrabble farms without off-farm income? Not really. They, too, had to work off the farm to make ends meet.

We've been realistic about having off-farm jobs to support the family farm. Without a well-paying off-farm job, we wouldn't have been able to stay small and still pay the mortgage and health care, not to mention kids' college and family vacations. Yet we didn't do off-farm work just for the money. Nancy liked her research and teaching role in environmental engineering at the university, and it's what brought us to Vermont in the first place. John's lacrosse coaching, teaching at the university, consulting, and other work have been personally and professionally enriching for him and the family.

Seeds of Self Reliance: A Farmer without Borders

After Peace Corps, John kept up his interest and enthusiasm for international exploration and languages (he is conversant in French, Spanish, and Haitian Creole) by participating in short-term international farmer-to-farmer consultancies dealing with sustainable agriculture. During these two- or three-week-long trips during the off-season, he focused on promoting ecologically based farming methods while Nancy held down the homefront. These programs were funded by USAID, international corporations, and nongovernmental aid organizations. On the surface they seemed mostly like public relations efforts with little long-term impact compared with the expenditures, but John took them seriously. He tried to connect to the international farmers on a personal level and help them figure out ways to remove obstacles to their goals. John has enjoyed these adventures and regarded them as opportunities for cross-cultural interactions and to share and exchange ideas. He has learned as much as, if not more than, the farmers he's worked with.

A trip to Haiti with two additional Vermont organic farmers brought a hand-powered chicken plucker to a local women's cooperative. The chicken plucker was designed and built by the father of John's friend David, who was also a participant in the project. With the new plucker, the women's cooperative was able to increase their chicken production and open a storefront in their town. The plucker design has since been copied and is helping other Haitian entrepreneurs.

FIGURE 12.8. David and John demonstrate the hand-cranked chicken plucker in Haiti. David's dad designed and built it.

Another farmer-to-farmer exchange took John to North India, where he worked with an Indian women's group on organic vegetable production. On that same trip, he met with Buddhist monks and visited their vegetable gardens. Among other things, the monks wanted to keep aphids off their tomatoes. John mentioned using a spray of soapy water as a possibility, but they quickly told him he had misunderstood. They didn't want to kill the aphids. That was against their worldview. Upon further discussion, it was agreed that the aphids were usually a problem for only a short while and only in the tomato patch. John was able to show them that the predatory ladybug and green lacewing larvae were most likely bringing the populations under control. They thought that was cool.

John also taught courses abroad through the University of Vermont Plant and Soil Science Department focused on community gardening and sustainable tropical agriculture. One Burlington College course in Cuba, co-taught with Will Raap, the founder of Gardeners Supply, compared and contrasted Vermont's food systems with Cuban food systems. Participating in these ongoing courses prompted us to start a complementary nonprofit called Seeds of Self Reliance. The main mission at that time was to promote gardens and farming as a way to build food security and community in Haitian *bateys* in the Dominican Republic and in schools and villages in Haiti. Students in the courses and workers in our nonprofit were excited to stay with, get to know, and roll up their sleeves to work alongside the Haitian community members.

Having a nonprofit gave us the ability to raise funds for seeds and tools and supported several groups of volunteers to live and work in these communities for three to six months. Our sustainable development work always focused on people-to-people contact and the exchange of ideas and information. We don't believe we are more knowledgeable about solving local problems than the people who live there, but we often have access to resources that can help remove obstacles the locals have identified. We might have new ideas to share, but so do they. We're all learning and sharing and believe these cross-cultural interactions are most important for world peace and getting to know others. This is an idea we've carried forward from our Peace Corps days.

In more recent years, we've been focusing our nonprofit work on local pollinator conservation projects that include school pollinator and edible gardens as well as bee education workshops. Promoting biodiversity and other ecological principles with other land managers is a continuing focus for us.

CHAPTER 13

bringing it home

There's a lot we can do on our own lands to enhance biodiversity that benefits people and wildlife. We've shared some of our approaches and ideas that others can adopt and adapt to meet their own needs. There are also community-based initiatives that can protect, conserve, and sustainably steward local resources. For example, supporting sustainable agriculture in our areas and working with local governments and conservation organizations can make a big difference in changing our unsustainable monoculture land management practices. We've had opportunities to participate in many of our local community-based initiatives, such as supporting and working toward conserving public lands, increasing native plantings in public areas, bike trail maintenance, and local presentations about biodiversity. It's been a great way to network with our local community members.

We also have strong ideas about the role that we can all play in education. We need to develop and nurture a cultural mind-set that keeps human beings within nature, not above or separate from it.

Ecoliteracy

Our definition of ecological literacy is the basic understanding and appreciation that everything (living and nonliving) is connected, and that we humans are animals and dependent on the Earth and the other creatures that inhabit it. It's an understanding that pollution and pesticides have consequences for other beings and ourselves even if we don't always understand the details of those impacts. It's an understanding of the importance of a biodiverse habitat and the problems with habitat destruction. Even if we live in a city and don't feel connected to trees, plants, and other animals besides humans, we are all dependent on the photosynthesis of plants for oxygen, food, and natural cycles.

We believe that a certain level of ecological literacy is paramount to the survival of our species. When Bill Clinton ran for president, one of his campaign slogans was "It's the economy, stupid!" We beg to differ. "It's the ecology!" If we don't protect and preserve our ecological underpinnings, there is no economy.

How do you stay upbeat and help people attain a basic level of ecological literacy? Ideally you start when kids are young, introducing them to the interconnectedness of all things. Sometimes you need to start with the warm and fuzzy animals, birds and mammals, or if insects, then butterflies and bumble bees. Butterflies seem so dainty and fairylike as they flit over the tall grass and sip nectar from the flowers. Reminding people that the caterpillars eating the leaves on their trees and shrubs are actually baby butterflies and moths can often cause an "Aha!" moment.

Bringing to people's attention that bumble bees and other bees pollinate our fruit and many of our vegetable crops also makes it easier for people to appreciate them. Bees are fuzzy and cute, which is a plus. Many people, though, confuse yellowjackets (in the wasp family) with bees. Yellowjackets can be much more aggressive than bees, especially when they want a sip of your sweet soda or a bite of your hamburger. People always want to share their sting stories. We warn them that swatting at wasps may get you stung, and it can hurt. So don't swat! A good life lesson.

A family member of ours who was afraid of "bees" recently visited us. She'd been stung recently, and her wrist had swelled up, with red streaks going up her arm. There are people allergic to bee venom who go into anaphylactic shock and can't breathe. If they don't get help within minutes, they can die. These cases are rare. The rest of us may have reactions that range from mild irritation to severe swelling like

bringing it home

FIGURE 13.1. Nancy created a Pollinators in Peril series of fiber art to bring awareness to the public about pollinator issues. Here, she stands with her exhibit of three pieces, *Got Milkweed?*, *Know Your Beekeeper*, and *Secret Toxic Garden*, at a NOFA Winter Conference.

our visitor. Usually, these aren't life threatening. A little information helped our visitor understand about "bee stings" on an intellectual level, but not necessarily emotionally. She was still afraid.

She was also confused about "bees" generally and specifically. Was it really a bee sting (from a honey bee or other bee)? Or was it from a yellowjacket or another type of wasp? It turns out it was a wasp, not a bee. It's good to know the difference.

We introduced her to honey bees by letting her look in an observation honey bee hive. We keep it in a building we use for nursery checkout and as a bee education center. We also showed her posters of native bees, and as we walked around the farm, we pointed out different examples of bumble bees and other native bees. She knew about the good things that bees do, but we also explained about the good things that wasps do. They're carnivores and use caterpillars and other insects as food for their young, making them great pest control for the garden and farm.

We couldn't get rid of the all the cultural conditioning and emotional baggage she carried against "bees," but by the end of the tour, she was differentiating between bees and wasps and realizing that they're not necessarily deadly or an insect she should really fear. She was able to get up close and personal with native bees on goldenrod and to start thinking of them as "pretty cute." Now if only this appreciation can withstand all the negative advertisements from pest control companies or alarmist news articles that she might see in the future.

Human beings are nature. Instead of a single being, the human body might be better thought of as an ecosystem. Adult humans have several pounds of microorganisms living in and on them. In fact, our prokaryotic cells (those cells lacking a nucleus, such as bacterial cells) are greater in number than the eukaryotic cells (those cells with a nucleus, such as our mammalian cells). The bacteria in our digestive system, often referred to as our gut biome, influence our immune system, food cravings, moods, our ability to learn, and much more. They also help us digest many plant foods that we might not otherwise be able to digest. We are only beginning to understand the importance of these symbiotic microorganisms of the human ecosystem, and how to better care for them and thus ourselves.

Being compassionate to other beings needs to extend to the other beings that coinhabit each of us. That starts with being good to our bodies, feeding ourselves healthy, nutritious food, avoiding toxins, and learning where and how to grow food that is healing to ourselves and the planet.

Limits to Growth

One of the most important lessons in ecology is that of carrying capacity, the maximum number of a species that can be sustained in a given ecosystem. Food, water, habitat, and other factors influence carrying capacity. When the population overshoots the carrying capacity, the population goes into decline, often quickly, as in a population crash. Depending on the species and the environment, overshooting might result in migration; death from disease, starvation, and predation; or a combination of these. There are limits to growth in every population, as witnessed often in nature with species population fluctuations.

Regarding humans, most of the current ecological thinking says that we are in overshoot. We are sustaining our current population and consumption rate because of our reliance on fossil fuel reserves and other resource-depleting practices (overfishing, unsustainable agriculture, and so on). Fossil fuels have allowed humans to produce energy-intensive fertilizers and energy-intensive irrigation methods used in unsustainable conventional food production. So far, it's worked, but we're losing arable lands, depleting aquifers that take thousands of years to recharge, and depleting our easily accessible oil reserves. Time is running out. *Limits to Growth: The 30-Year Update* by Donella Meadows, Jorgen Randers, and Dennis Meadows is a great book for those interested in learning more about this important and urgent topic.

As we attempt to personalize the concepts of overshoot and limits to growth, we've been drawn to ideas of simplicity and degrowth. The book *Voluntary Simplicity* by Duane Elgin first came out in 1981, recommending that each of us personally take a less harried and less consumptive approach to living on this planet. By doing so, we take personal responsibility for our own health and the health of the planet and can at least achieve empowerment and spiritual growth through our actions. We are not naïve about the challenges ahead and the broad structural changes that need to occur to slow society's self-destructive trajectory and reduce current and future suffering of all beings. However, we find glimmers of hope in the positive power of regenerative land stewardship and the strengthening of relationships with our communities.

Degrowth (*décroissance* in French) is a philosophy that encourages downscaling of production and consumption. This has become a political, economic, and social movement in many places and argues that environmental and social problems result from capitalist overconsumption economics. *The Power of Restraint* by French Algerian farmer Pierre Rabhi was an inspirational book on this topic for us. Simplicity and degrowth ideas are not about living in poverty or becoming martyrs. Both emphasize consuming less and devoting more time to family, community, nature, and the arts. In our society, it's easy to get caught up in doing and having more, but by investigating simplicity and degrowth in our own lives, we often realize we don't need more things or more money. What we really need is more time to enjoy the simple pleasures, a wholesome meal with friends and family, a baby's smile, and the sweet smell of plum blossoms. And to slow down and breathe deeply.

Learning to Love the Insects

The story goes that the Dalai Lama was once asked how to change the world. His reply was to teach children to love insects. This beautiful sentiment of teaching love for the small life-forms that coinhabit the planet with us is something that we all can learn. Or if not love, then we can at least learn appreciation, compassion, and respect. It's not hard to do. It might even spill over to loving each other more too.

There's a warm spot in our hearts for insects, for many reasons. We first got to know each other in an entomology class in college in 1978. For that class, each student was required to make a large insect reference collection with over three hundred specimens from

FIGURE 13.2. Nolan holding a male eastern dobsonfly. The males look menacing with their elongated mandibles but are harmless. The females can bite, however, as can the immature aquatic stage, called hellgrammites; top invertebrate predators in rocky streams.

different taxonomic Orders and Families. One of our first conversations revolved around the fact that Nancy had cockroaches in her college apartment (another taxonomic Family!) and John had silverfish (another taxonomic Order!). Trades were made, and the rest is history. That class also led us to delve into the amazing diversity, ecological importance, and transformational life cycles of insects. That introduction had a big impact on our compassion and fostered the positive relationship we have with them.

Insects have gotten a bad rap over the years, mostly from chemical and other industries trying to make a buck off people's fear. We often wonder why so many people resort to chemicals as their first reaction to insect matters. In our Western culture, we've developed such a misplaced fear and loathing of these little critters when it really should be a fear of the pesticides and chemicals getting doused on our food and in the environment. Insects are amazing creatures. Their species diversity and sheer numbers on the planet are mind-boggling. And generally, plants can handle a certain amount of leaf damage. Don't panic. It's better to learn to live with them.

The current lack of ecological understanding within the general public and among government officials concerning the important role of insects in local and global ecosystems is alarming. The war on insects that is being waged in homes, gardens, and on farms reflects that ignorance. This war encourages a mentality of fear. The desire to control people, animals, and the environment is really based on human insecurities and a fear of change. The inability to face the fact that life means change and impermanence prevents many people from truly enjoying life. Most people live so much in the past or the future, instead of the here and now. It's a challenge, but perhaps if we increase our learning and appreciation for the natural world, observe its natural cycles of life, death, decay, and regeneration, and learn to love the insects, we could all be happier and healthier.

Will Allen in *The War on Bugs* and Joanna Louke in *The Voice of the Infinite in the Small* convincingly argue that Western society's promotion of fear and loathing around insects started in the United States only a little over a hundred years ago, which is not so very long ago in human history. With their six legs, big eyes, ability to fly, and exotic sounds, insects can seem pretty foreign, though. They fly at us and crawl over us; a few bite and make our skin welt up; flies buzz our ears; and a few even carry disease. How can we possibly love them? Joanna Louke points out that in other cultures around the world, insects have been revered as messengers and guides, and as deities incarnate.

Our own experience in Africa, where the local people weren't afraid of them or trying to kill them all the time, even occasionally using them as a food source, brought us face-to-face with our own cultural conditioning. While the Kenyans and Malians we worked with didn't want various insects eating their crops and didn't like the mosquitoes that carried malaria, it was clear that they didn't have the attitude of fear and hatred we've seen in the United States, which is only getting worse as people become more disconnected from the natural world. While a few insects (and other arthropods) do carry disease, and are therefore cause for concern, we think as a society we need to build more compassion and appreciation of nature, including these small creatures. That's one thing increased ecoliteracy can do.

Many people are already used to living with nonhuman companions. Dogs and cats roam the house and lie on the furniture. They're often dirty and smelly, but we love them. We find it distressing that many people use insecticidal powders, sprays, and collars on their dogs and cats as a prophylactic against fleas and ticks and then let their kids snuggle and sleep with these pesticide-laden furry friends. This is another good example of our fear and loathing of insects, and our indifference to the chemicals, which are really what we should be afraid of. In our twenty-six years of owning dogs and cats, we've had only a few flea outbreaks. That's when we've given them a treatment. We do tick checks on the dogs and cats at the same time we do tick checks on us. No need to poison them or us.

We liked watching ants as kids, dropping cookie crumbs and seeing them try to pick one up to carry the crumb back to their nest. What a feat! We have a lot of ant mounds out in the back meadow of the pollinator sanctuary, big hummocks visible in the winter and early spring before the grass grows and hides them. We've identified them as the Allegheny mound ant. The workers tend aphids, treehoppers, and other insects for honeydew. The ants also collect plant saps and nectar. Their mounds are used by other insects such as beetles, flies, and caterpillars. The ants protect the caterpillars in the nest from predators and the caterpillars feed the ants with a sugary liquid that they secrete. Wow!

Ants are true social insects with a division of labor that relies on one (or only a few) reproductive females, which is truly an evolutionary feat. It's estimated that ants constitute one-third of all insect mass on Earth. They, not earthworms, are the major movers and turners of soil and are nature's recyclers of plant and animal debris. They also serve as food for spiders, birds, and other insects.

FIGURE 13.3. A monarch butterfly sips nectar from ironweed, a late-season blooming flower in the pollinator sanctuary.

We also have carpenter ants in and around the old farmhouse. They don't actually eat wood but clean out wet and rotting wood to build their living quarters. Once when we had new windows installed in the farmhouse, the contractor found a rotten sill and rotten rafters that were also occupied by carpenter ants. He had to replace a lot of rotten wood. That's the problem with old houses and old buildings: whenever you do work on them, you find out three or four more things that need to be fixed as well. Sometimes ignorance is bliss.

Every day we are humbled by and in awe of the complexity and beauty that surround us, from the small beetle on a leaf to the surrounding forested hills. We don't understand how people can love the monarchs and milkweed yet hate the beautiful yellow oleander aphids or tussock moth caterpillars that also live on the milkweed. Or how it is we've learned to hate the codling moth larvae (worm in the apple) that have coevolved and adapted to love the same fruit we do. We are all different expressions of life on this planet. It doesn't have to be a war. We don't want to inadvertently bite the worm in the apple, but we don't want to hate it either.

Small Things

History has borne out the power and importance of small farms and home gardens in providing nutritious food for communities, especially in times of stress. While we'd like to see the current monoculture farms become biodiverse regenerative systems, we think it's better to have more small-scale regenerative practitioners than just a few large-scale ones. Small farms can be the economic engines in rural environments and revitalize local economies. Money and good intentions circulate in our small towns like nutrients in a farm system. Buying from the local hardware and feed stores, supporting other entrepreneurs, wholesaling to local grocers, and supporting local employees are all ways farms close the money loop and regenerate rural communities. Sharing, teaching, and building alliances also create social networks and community. Producing healthy food, purpose and meaning, and a connection to the natural world is a priceless contribution that biodiverse small farms and gardens can make to our communities, no matter where they are.

One of the speakers at a recent talk we attended about surviving the future said that "the next big thing is a bunch of small things." Small acts creating pockets of persistence and resistance to environmental and social degradation multiplied by millions of people will fuel the revolution we need to heal the planet and ourselves. We can all effect change.

APPENDIX

common names to scientific names

The following list includes important plants and arthropods mentioned in the text of this book.

Agastache	*Agastache foeniculum*	Basswood	*Tilia americana*
Ajuga	*Ajuga* spp.	Bayberry	*Myrica* spp.
Alkanet	*Alkanna tinctoria*	Beach plum	*Prunus maritima*
Alfalfa	*Medicago sativa*	Bee balm	*Monarda* spp.
Allegheny mound ant	*Formica exsectoides*	Bishop's-weed	*Aegopodium podagraria L.*
American chestnut	*Castanea dentata*	Bistort superba	*Persicaria bistorta 'superba'*
American elm	*Ulmus americana*	Black cherry	*Prunus serotina*
American witch hazel	*Hamamelis virginiana*	Black cohosh	*Actaea racemosa*
Apple	*Malus domestica*	Black currant	*Ribes nigrum*
Apple maggot	*Rhagoletis pomonella*	Black locust	*Robinia pseudoacacia*
Aronia	*Aronia melanocarpa*	Black walnut	*Julgans nigra*
Ash, mountain	*Sorbus americana*	Black-eyed Susan	*Rudbeckia hirta*
Ash, white	*Fraxinus americana*	Blazing star	*Liatris spicata*
Aster	*Aster* spp.	Bleeding heart	*Lamprocapnos spectabilis*
Baptisia	*Baptisia australis*	Blue azure butterfly	*Celastrina neglecta*

Blue orchard bee	*Osmia lignaria*
Blue vervain	*Verbena Hastata L.*
Blueberry	*Vaccinium cyanococcus*
Bluejoint reedgrass	*Calamagrostis canadensis*
Boneset	*Eupatorium perfoliatum*
Box elder	*Acer negundo*
Buckwheat	*Fagopyrum esculentum*
Bumble bee	*Bombus* spp.
Burdock	*Arctium minus*
Burr oak	*Quercas macrocarpa*
Buttercup	*Ranunculus* spp.
Butterfly weed	*Asclepias tuberosa*
Buttonbush	*Cephalanthus occidentalis*
Catnip	*Nepeta* spp.
Centaurea	*Centaurea* spp.
Cherry tomato	*Solanum lycopersicum* var. *Cerasiforme*
Chinese chestnut	*Castanea mollissima*
Chokecherry	*Prunus virginiana*
Chive	*Allium schoenoprasum*
Clove currant	*Ribes odoratum*
Clover	*Trifolium* spp.
Cockroach	*Blattella germanica*
Codling moth	*Cydia pomonella*
Comfrey	*Symphytum officinale*
Common hackberry	*Celtis occidentalis*
Common whitetail dragonfly	*Plathemis lydia*
Concord grape	*Vitis labrusca*
Corn	*Zea mays*
Creeping Charlie	*Glechoma hederacea*
Creeping phlox	*Phlox subulata*
Crocus	*Crocus sativus*
Daffodil	*Narcissus pseudonarcissus*
Dandelion	*Taraxacum officinale*
Digger bee	*Anthrophora* spp.
Dogwood, Cornelian cherry	*Cornus mas*
Dogwood, pagoda	*Cornus alternifolia*
Dogwood, red osier	*Cornus sericea*
Dogwood, yellow-twig	*Cornus sericea Flaviramea*
Dogwood, silky	*Cornus amomum*
Early Gold pear	*Pyrus ussuriensis*
Eastern dobsonfly	*Corydalus cornutus*
Echinacea	*Echinacea purpurea*
Elderberry	*Sambucus nigra*
Elderberry (American subspecies)	*Sambucus nigra canadensis*
English walnut	*Juglans regia*
European apple sawfly	*Hoplocampa testudinea*
Filbert	*Corylus maxima*
Fir	*Abies* spp.
Flat-topped aster	*Doellingeria umbellata*
Flemish Beauty pear	*Pyrus communis*
Forget-me-not	*Myosotis scorpioides*
Fowl bluegrass	*Poa palustris*

Appendix

Fritillary butterfly	*Speyeria cybele*	Joe-pye weed	*Eutrochium purpureum*
Globe thistle	*Echinops* spp.	Juneberry	*Amelanchier canadensis*
Gogi berry	*Lycium barbarum*	Labrador tea	*Rhododendron tomentosum*
Goldenrod	*Solidago* spp.	Ladybird beetle (ladybug)	Coccinelidae
Gooseberry	*Ribes uvacrispa*	Lady's mantle	*Alchemilla* spp.
Gooseberry sawfly	*Nematus ribesii*	Large milkweed bug	*Oncopeltus fasciatus*
Grass-carrying wasp	*Isodontia* spp.	Leaf-cutting bee	Megachilidae
Gray birch	*Betula populifolia*	Lilac	*Syringa vulgaris*
Green darner dragonfly	*Anax junius*	Lily of the valley	*Convallaria majalis*
Ground beetle	*Caribidae* spp.	Lingonberry	*Vaccinium vitis-idaea*
Hairy vetch	*Vicia villosa*	Littleleaf linden	*Tilia cordata*
Hardy kiwi	*Actinidia arguta*	Lungwort	*Pulmonaria* spp.
Hawthorn	*Crataegus* spp.	Maple	*Acer* spp.
Hazelbert	*Corylus* spp.	Marsh reedgrass	*Calamagrostis Canadensis*
Hemlock	*Tsuga canadensis*	Mason bee	*Osmia* spp.
Hemp	*Cannabis sativa*	Mason wasp	Eumeninae
Highbush cranberry	*Viburnum trilobum*	Maximilian sunflower	*Helianthus maximiliani*
Honey bee	*Apis mellifera*	Meadow rue	*Thalictrum* spp.
Honey locust	*Gleditsia triacanthos*	Milkweed, common	*Asclepias syriaca*
Honeyberry	*Lonicera caerulea*	Milkweed leaf beetle	*Labidomera clivicollis*
Honeysuckle	*Lonicera periclymenum*	Milkweed tussock moth	*Euchaetes egle*
Hummingbird, Ruby-throated	*Archilochus colubris*	Mock orange	*Philadelphus lewisii*
Iris	*Iris* spp.	Monarch butterfly	*Danaus plexippus*
Plasterer bee	*Colletes* spp.	Mountain ash	*Sorbus aucuparia*
Jacob's ladder	*Polemonium caeruleum*	Nanking cherry	*Prunus tomentosa*
Japanese beetle	*Popillia japonica* Newman	Nannyberry	*Viburnum lentago*
Jewelweed	*Impatiens capensis*		

Narcissus	*Narcissus* spp.	Radish	*Raphanus raphanistrum sativus*
Nasturtium	*Tropaeolum majus*	Raspberry	*Rubus idaeus*
New York ironweed	*Vernonia noveboracensis*	Red currant	*Ribes rubrum*
Ninebark	*Physocarpus opulifolius*	Red meadowhawk	*Sympetrum* spp.
		Red milkweed beetle	*Tetraopes tetrophthalmus*
Nodding sedge	*Carex gynandra*	Red osier dogwood	*Cornus sericea*
Oats	*Avena sativa*	Red-humped caterpillar	*Schizura concinna*
Oregano	*Origanum vulgare*		
Osage orange	*Malcura pomifera*	Reed canary grass	*Phalaris arundinacea*
Ostrich fern	*Matteuccia struthiopteris*	Rhubarb	*Rheum rhabarbarum*
Pea	*Pisum sativum*	Root rot fungus	*Armillaria mellea*
Pear	*Pyrus* spp.	Russian mulberry	*Morus alba*
Penstemon	*Penstemon* spp.	Rusty patched bumble bee	*Bombus affinis*
Peony	*Paeonia* spp.	Rye	*Secale cereale*
Peppermint	*Mentha piperita*	Saint-John's-wort	*Hypericum perforatum*
Phacelia	*Phacelia tanacetifolia*	Saskatoon berry	*Amelanchier alnifolia*
Phlox, wild	*Phlox* spp.	Sea kale	*Crambe maritima*
Phragmite	*Phragmites* spp.	Seaberry / Sea buckthorn	*Hippophae rhamnoides*
Plantain	*Plantago major*		
Plasterer bee	*Colletes* spp.	Siberian peashrub	*Caragana arborescens*
Plum	*Prunus americana*	Siberian squill	*Scilla siberica*
Plum curculio	*Conorachelus nenuphar*	Silkworm	*Bombyx mori*
		Silky dogwood	*Cornus amomum*
Poverty oatgrass	*Danthonia spicata*	Silver maple	*Acer saccharinum*
Primrose	*Primula vulgaris*	Silverfish	*Lepisma saccharina*
Pumpkin	*Cucurbita*	Small carpenter bee	*Ceratina* spp.
Purple osier willow	*Salix purpurea*	Small milkweed bug	*Lygaeus kalmia*
Pussy willow	*Salix discolor*	Sneezeweed	*Helenium autumnale*
Queen Anne's lace	*Daucus carota*	Sow bug	*Oniscidea* spp.

Appendix

Soybean	*Glycine max*	Tomato	*Solanum lycopersicum*
Speckled alder	*Alnus incana*	Tracheal mite	*Acarapis woodi*
Spotted wing drosophila	*Drosophila suzukii*	Tree hopper	Membracidae
Springtail	*Collembola*	Tricolored bumblebee	*Bombus ternarius*
Squash bee	*Peponapis pruinosa*	Upland bentgrass	*Agrostis perennans*
Stinkbug	Pentatomidae	*Varroa* mite	*Verroa destructor*
Strawberry	*Fragaria* spp.	Violet	*Viola* spp.
Sugar maple	*Acer saccharum*	Virginia rose	*Rosa virginiana*
Sumac, staghorn	*Rhus typhina*	Virginia wildrye	*Elymus virginicus*
Summersweet clethra	*Clethra alnifolia*	Wasp	Apocrita
		Wetland rose	*Rosa palustris*
Sunchoke (Jerusalem artichoke)	*Helianthus tuberosus*	White oak	*Quercus alba*
Sunflower	*Helianthus annuus*	White pine	*Pinus strobus*
Sunflower bee	*Svastra obliqua*	Wild parsnip	*Pastinaca sativa L.*
Swamp milkweed	*Asclepias incarnata*	Winterberry holly	*Ilex verticillata*
Swamp white oak	*Quercus bicolor*	Wintergreen	*Gaultheria procumbens*
Sweat bee	Halictidae	Yellowjacket	*Vespula maculifrons*
Tachinid fly	*Istocheta aldrichi*	Yellow loosestrife	*Lysimachia vulgaris*
Tart cherry	*Prunus mahaleb; Prunus cerasus*	Yellow oleander aphid	*Aphis nerii*
Tick	*Ixodida scapularis*	Upland bentgrass	*Agrostis perennans*

Index

Note: Page numbers in *italics* refer to figures, photos, and illustrations. Page numbers followed by *t* refer to tables.

A

Abenaki people, 3
Adams elderberries, 140, 142
adaptations for climate change.
  See climate change adaptations
advertising, 227
agritourism programs, 213–15, *214, 216–17*
agroecology, 44–45, 46
Agroecology Livelihoods Collaborative, 45
agroforestry, 95–108
  establishment and care of perennial plantings, 105–8, *108*
  overview, 95–96, *96*
  perennial vegetables, 103–5, *104*
  polyculture fruit trees and guild plants, 96–99, *98*
  resilience and, 46
  silvopasture and alley cropping, 99–103, *100, 101, 102, 103*
Alderman plums, *113*
Alisha (farm worker), 132, *140*
Allegheny mound ants, 238
Allen, Will, 237
alley cropping
  Knoll Orchard, 99, 100–103, *100*
  milkweed stands, 167
  no-till practices, 80
  overview, 15
  plantings for mulch, 84

American basswood.
  See basswood
American chestnut, 127
American-Chinese hybrid chestnut, 127
American elm, 158–59, 162
American plums
  growing information, 124, *124*
  in the pepinyè garden, 153
  in the pollinator sanctuary, 156
  rootstock for, 115
American witch hazel, 155–56
ammonia, as food for nitrifiers, 62, 63
ant mounds, *176*, 238
Antonovka rootstock, 97, 115
apple cider, 51, 97, 222–24, *223*
apples
  abandoned orchards, 118, 222
  favorite varieties, 119*t*
  growing information, 116–120, *116, 118*
  in hoophouses, 14, *90*, 91, *93*, 94
  no-till production in alleys, 80
  organic approach to, 51, 116–17, 223
  pest management, 186–87, *186*
  rootstock for, 97, 115

  See also Front Lawn Orchard; Knoll Orchard
apple scab, 117, 118, 187
apricots
  growing information, 124
  in hoophouses, 14, *93*
  self-fruitfulness of, 122
Araucana laying hens, 24
Armillaria mellea (root rot), 125
aronia
  field 6 plantings, 76
  as lawn replacement, *54*
  overview, 138–142, *138*
  propagation of, 148–49
  wholesale markets, 226
ash trees, 174
asparagus, 103, *104*
Aurora honeyberries, 144

B

back driveway views, *8*
back pasture, 14, 95, 151
  See also pollinator sanctuary
Bailey, Janet, 30
Bailey, Jay, 30
bark-peeling camps, 228
barn
  early years, 6–7
  flooding of, *72*, 74–75
  roof water collection, 87
  wintertime food waste composting project, 68–69

basswood
- growing information, 159, 162
- Japanese beetle leaf damage, 180
- pollen and nectar from, 205
- streamside location, 174

bayberry plantings, 99
beach plums, *12–13*, 144–46, *145*
bee boxes, 200–202, *201*
beekeeping
- bumble bees, 209–10, *210*
- honey bees, 206–9

bees, 193–210
- beekeeping, 206–10, *210*
- ecoliteracy about, 232–34
- floral resources for, 203–6, 204t, *206*
- honey bees, 194–95, 206–9, 232–34
- overview, 193–94, *194*
- in the pepinyè garden, 153, *154*
- pesticide-free approach, 195–97, *196*
- solitary nesting bees, 200–203, *201*, *202*
- squash and sunflower bees, 102, 197, 200, 202
- stresses on, 194–95
- *See also* bumble bees

beetles, Japanese, 179–181, *180*
bench grafting, 115
Ben Lomand black currant, 131
Ben Sarek black currant, 131
Berry Blue (Czech 17) honeyberries, 144
berry bushes
- deer protection for, 106–7
- first plantings, 14
- nylon footies for protection, *185*
- picking hygiene for managing spotted wing drosophila infestations, 184–85
- *See also specific types*

Betty (family cow), 26

Beyond the War on Invasive Species (Orion), 183
"biggering," 221
Big Picture Project, *212*
biodiversity
- apple trees, 97
- benefits of edges for, 157, 175
- biological inputs for, 45
- coexistence with pests, 190–91
- farm's focus on, 7
- within fields, 105
- of guild plants, 99
- in healthy soil, 57, 62–63
- importance of, 6, 49–50
- multifunctional beauty of, *8*
- native plantings, 152
- in the pepinyè garden, 152, 156
- pest management benefits, 177–79, 180–81, 187
- pollinator benefits, 193, 230
- role in resilience, 38

biodynamics, 46–47
biological indicators of soil health, 59t, 60–61
birds
- cleanup of crops by, 178
- netting for protection against, 134, 143, 178
- in the pepinyè garden, 152–54
- pest management benefits, 187
- in the pollinator sanctuary, 157, 175
- return of, through rewilding endeavors, 49–50, *50*
- *See also specific types*

bishop's-weed, 191–92
black cherries, 156, 174
black currants
- first plantings, 14, 48–49, *48*
- growing information, *128*, 129–132
- in Knoll Orchard, 98
- in the pollinator sanctuary, 176

raised bed plantings, 78
steam juicing, 219, 220
wholesale markets, 226

black-eyed Susan, 165
black locust
- chop and drop system for branches, 84
- pollen and nectar from, 205
- in the pollinator sanctuary, 156, 160–61, 162–63

Black Velvet gooseberries, 137
black walnut, 126–27
black willow, 163–64, 203
blemished apples, 51, 117–18
blister rust, 130–31
blueberries
- early plantings, 129
- fall foliage, *36*
- overview, 11
- pollen and nectar from, 203–4
- propagation of, 149
- young plants, *12–13*

blue orchard bees, 200, 201
Bob Gordon elderberries, 140, *140*, 142
Borealis honeyberries, 144
bottle-fed lambs, *18*
box elder
- streamside location, 50, 203
- woodchips from, 81, 84

Boyden Valley Winery, 48
breweries, wholesale markets for fruit, 14, 130, 226
Bringing Nature Home (Tallamy), 152, 225
bucket-training of animals, 28
buckwheat, for cover cropping, 70
Bud 118 rootstock, 97
Bud 9 rootstock, 115
bud grafting, 116
bumble bees
- attraction to honeyberries, 143
- commercial beekeeping, 209–10, *210*
- ecoliteracy about, 232–34

248

Index

floral resources for, 203–4
in hoophouses, 89, 91
nesting needs, 197–99, *198*, *199*
in the pepinyè garden, 154
bumblekulturs, 9, 86, 199, *199*
Burlington Farmers Market, 220, *220*, 222, 223–24
burr oak, 159
butterflies
    monarch butterflies, 167–170, *168*, *239*
    in the pollinator sanctuary, 164, 175
    viceroy butterflies, 169
buttonbush flowers, 154–55, *154*
buzz pollination, 91, 203

C

CAFOs (confined animal feeding operations), 29
carbon cycle, 61
carbon sequestration
    methods for, 58
    with no-till systems, 63
    regenerative agriculture practices for, 44
    woody shrubs and trees, 15
carbon-to-nitrogen ratios
    for composting, 64, 66
    woodchips, 81
cardboard mulch, 79, 80–81, *81*
Carpathian walnut trees, 127
carpenter ants, 239
carpenter bees, 202, *202*
carrying capacity, 234–35
catbirds, 153–54
cavity-nesting bees, *196*, 200–203, *201*, *202*
cedar waxwings
    aronia consumption, 142
    cranberry consumption, 154
    honeyberry consumption, 143, *144*
    red currant consumption, 134, *135*, 189
    in the riparian zone, *96*
    winterberry holly consumption, 166

cellophane bees, 202
chemical indicators of soil health, 59*t*, 60
cherries
    in hoophouses, 14
    rootstock for, 115
    sweet, 122
    tart, 115, 121, 122
cherry tomatoes, growing in hoophouses, 49, 88, 91
chestnut, 127
chicken plucker, 229, *229*
chickens
    composting of manure and food scrap leftovers, 68–69
    laying hens, 24, *25*, 68–69
    for meat, 22, *23*, 33
chicken tractors, 22, *23*
Chinese chestnut, 127
chop and drop system, 84, 98
ciders
    apple, 51, 97, 120, 222–24, *223*
    fruit, 222–24, *223*
    hard, 120
    Wild Plum Cider, 146
classroom visits to the farm, 211–13, *212*
cleft grafting, 115–16
climate change
    changeover in farm's focus due to, 7
    effects in Vermont, 91–92, 159
    effects on the farm, 114
    regenerative agriculture benefits, 44
climate change adaptations, 73–94
    dry-weather watering, 86–87, *87*
    flooding, *72*, 74–76, *76*
    hoophouses, repurposing of, 88–94, *89*, *90*, *93*
    hügelkultur practices, 85–86, *85*
    mulching practices, 79, 80–85, *80*, *81*, *82–83*

    overview, 73–74
    regenerative no-till soil practices, 78–80, *79*
    riparian zones, 76–77, *82–83*
    wet, heavy soils, 77–78
"Close the Loop Chicken Coop," 68–69
clove currants, 132–33, *133*
codling moth caterpillars, 187, 239
cold stratification, 170
Coleman, Eliot, 19
colony collapse disorder, 208
comfrey
    for mulch, 84
    in polycultures, 99
community-based ecological initiatives, 231
composting
    carbon-to-nitrogen ratios, 64, 66
    importance to healthy soil, 63–69, *65*, *67*
    location for, 9, *11*
Concord grapes, *87*
confined animal feeding operations (CAFOs), 29
conservation plants, 7, 149, 225
    See also specific types
Consort black currant, 131
coppicing, 81, 84
Cornell University soil assessment, 58–60, 59*t*
Cornish/Barred Rock cross cockerels, 23, *23*
Cornish/White Rock cross cockerels, 23
cover crops
    floral resources for bees from, 205, *206*
    hogging down, *27*
    role in soil health, 69–71, *71*
CSA (community-supported agriculture)
    early years, 33, *35*
    meat CSA, 19, 26, 33
Cuba, farmer-to-farmer exchanges, 230

Cuckoo Marans laying hens, 24
currants
　clove currants, 132–33, *133*
　farmers market sales, *220*
　favorite varieties, 136*t*
　propagation of, 146, 148
　red currants, 134–35, *134*, 188–191, *189*
　See also black currants
currant sawfly. *See* gooseberry sawfly
Czech 17 (Berry Blue) honeyberries, 144

D

dairy farming, 26–27
David (farmer), 33, 229, *229*
decomposition
　of cover crops, 69–70
　of plant residues, 62
deer protection, *104*, 106–7
degrowth, 235–36
demonstration garden. *See* pepinyè garden
denitrifiers, 63
disease-resistant varieties
　apples, 119*t*, 186–87
　for integrated pest management, 117
diseases
　apples, 118, 187
　ash trees, 174
　elm trees, 158
　mulberry trees, 125
　Ribes spp., 130–31
　stone fruits, 122
diversity equals stability tenet, 19, 21
dobsonflies, *236*
dogwood plantings
　hardwood cuttings of, 99, *146*
　in the pepinyè garden, 154, *155*
　in the pollinator sanctuary, 156
draft horses, 30–32, *32*, 64
dry-weather watering, 86–87, *87*
Dudley (stud boar), 26

Dutch elm disease, 158
dwarf rootstocks, 94, 115

E

ecoliteracy, 232–34, *233*
ecological understanding and education, 231–240
　benefits of walking the land, 1
　college studies, 4
　ecoliteracy, 232–34, *233*
　farm's focus on, 7
　interconnectedness of all things, 1, 232
　limits to growth, 234–36
　love for insects, 236–39, *236*, *239*
　power of small things, 240
　resilience and, 38–39
　student programs at the farm, 211–13, *212*
　walking the land, 1
economic considerations
　benefits of small farms, 240
　"biggering" cautions, 221
　fruit nursery, 224–25
　importance of flexibility, 226–27
　off-farm income, 227–28
　resilience strategies, 41*t*
　simplicity and degrowth, 235–36
　social and ecological connections, 39
　unsustainability of the early years, 33–34, 37–38
　wholesale markets, 48, 226
ecosystems
　changing nature of, 157
　ecological thinking in managing pests, 183
　factors in supporting a honey bee colony, 209
　human body as, 234
　interrelationships in, 168
　limits to growth, 234–36
edges
　in the pollinator sanctuary, 157, 172–74, *173*, *175*

species for, 49, *52–53*
education. *See* ecological understanding and education
eggmobiles, 24, *25*
elderberries
　deer protection for, 106–7
　in field 6, 76, *77*
　fruit syrups of, 219, 220, *222*
　hygiene for managing spotted wing drosophila infestations, 184, 185–86, *185*
　landscape fabric mulch for, *80*, *82–83*
　overview, 138–142, *140*, 141*t*
　in the pepinyè garden, 154–55
　in the pollinator sanctuary, 163–64, *165*
　propagation of, 146, 148
　steam juicing, 219, *220*
　in wet, heavy soils, 78
　wholesale markets, 226
elderflowers, 138–39, 141, 155, *155*
electric fencing, *16*, 24, *26*
Elgin, Duane, 235
Elm Grove Farm (former name of farm), 158
elm trees, 158–59, 162
emerald ash borer, 174
Entrust (spinosad), 190
erosion
　cover-cropping benefits, 69
　grassy waterway benefits, 77
　riparian zone restoration benefits, 77
　from tillage, 63
European apple sawfly, 187
European elderberries, 165
European plums, 122

F

factory farms, 29
Fair Winds Farm, 30
false scorpions, 67–68
family cows, 26–27
The Farm Between
　changing focus of, 7

Index

diversity of products from, 218
goals for, 18–19
holistic resource management approach, 19–21
map of, *10*
name choice for farm, 218, 221
need for changes after the early years, 32–35, 37–38
stressors and corresponding resilience strategies, 40–41*t*
timeline of, 20*t*
See also specific areas of the farm; specific crops; specific farming practices
farmer-to-farmer exchanges, 87, 228–230, *229*
feeder pigs, 19, 26
fences and screens
electric fencing, *16*, *24*, 26
for pest protection, *104*, 106–7
feral honey bees, 208
field 3
hügelkultur, *85*
mixture of plants in, 176
raised beds, 78
field 6
flooding of, *72*, *74*, 75–76
switch to perennial plantings, 76
field trips to the farm, 211–13, *212*
fire blight, 117, 187
Flemish Beauty pear, 121
flexibility, need for, 226–27
flooding, climate change adaptations, *72*, 74–76, *76*
floral resources for pollinator, 203–6, 204*t*, *206*
food scraps, composting via chickens, 68–69
food sovereignty, 45
Freckles (ewe), 214
fritillary butterflies, 175
frogs, in the pollinator sanctuary, 175

Front Lawn Orchard
apple trees, 97, *112*, 116, 186–87
development of, 51, *54*
laying hens, *25*
map of, *11*
mowing, 119–120
soil of, 113
time to production, 109
young trees in bloom, *110–11*
frost concerns, stone fruits, 122
fruit and nut trees, 109–27
apple trees overview, 116–120, *116*, *118*, 119*t*
basic requirements, 112–14
Front Lawn Orchard, *110–11*, *112*
gradual approach to planting, 109
grafting overview, 114–16, *116*
mulberry trees overview, 125
nut trees overview, 125–27, *126*
pear trees overview, 120–21
stone fruits overview, 121–24, 123*t*, *124*
time to production, 109, 112
See also specific crops
fruit ciders, 222–24, *223*
fruit nursery, 224–25, *225*
fruit production. *See* perennial fruit production; uncommon fruits; *specific crops*
fruit syrups and snow cones, 218–221, *219*, *220*
fruit trees
deer protection for, *104*, 106
in hoophouses, 14, 91–92, 94
See also specific crops
fungal-dominated compost, 66
fungi, importance to healthy soil, 57–58

G

Gaia's Garden: A Guide to Home-Scale Permaculture (Hemenway), 45

geological features of the farm, 3
Global Assessment of Land Degradation and Improvement, 56
Gloucester Old Spot-cross sows, 26
glyphosate (Roundup), 63
Gnarly Fruit Collective, 222
gogi berries, 147*t*
goldenrod, in the pollinator sanctuary, 171
gooseberries
farmers market sales, *220*
favorite varieties, 136*t*
overview, 136–38
propagation of, 146, 148
raised bed plantings, 78
gooseberry sawfly, 188–191, *189*
gout weed (bishop's-weed), 191–92
grafting, 114–16, *116*
grain buckets, 26, 28
grapes, *87*
grass-carrying wasps, 201–2, *201*
grass-fed meat, benefits of, 29, 30
grassy areas
native grass plantings, 171–72
reclaiming of, 163–65
by waterways, 77
See also pasture
grazing, 21, 70–71, *71*
See also management intensive grazing
Green Mountain Draft Horse Association, 30
ground elder (bishop's-weed), 191–92
ground-nesting bees, 200, 202
growth, limits to, 234–36
guilds, in the orchards, 98–99

H

hackberry, 165
Haiti, farmer-to-farmer exchanges, 229, *229*, 230
halal slaughtering of animals, 29

hard cider, 120
hardwood cuttings, 146, *146*, 148
hardy kiwi, 147*t*
harvesting fruit
    black currants, 131–32
    clove currants, 133
    elderberries, 139, 141
    gooseberries, 137
    honeyberries, 143–44
    hygiene for managing pests, 184–85, 187
    red currants, 135
Harvey (calf), 27
haskaps. *See* honeyberries
Hayden, Connor, *18*, 74
Hayden, John P.
    coaching work, 64, 228
    college studies, 4, 236–37
    consulting and teaching work, 5, 34, 228–230
    draft horse training, 30
    farmer-to-farmer exchanges, 212, 228–230, *229*
    field work, *32*
    graduate studies, 5, 51, 116
    Irish heritage, 47, 166
    Peace Corps service, 4–5, 238
    permaculture class, 45
    purchase of farm, 5
    Salatin, Joel visit, 19
    scything, *108*
Hayden, Nancy J.
    art pursuits, 34–35
    college studies, 4, 236–37
    fiber arts work, 171, *233*
    graduate studies, 5
    Peace Corps service, 4–5, 238
    purchase of farm, 5
    raspberry picking, *89*
    Scandinavian heritage, 47
    university position, 35, 228
Hayden, Nolan, *18*, 74, *236*
haying, 14, 21–22
hazelbert trees, 14, 125–26, *126*
hedgerows, 46, 95–96, 126, 164
Hemenway, Toby, 45

hemp
    alley cropping, 102, *102*, 103
    no-till production, 80
heritage apple orchard, 15
highbush cranberries, 154, 163–64
high-temperature composting, 64–66, *65*
hogging down of cover crops, *27*
holistic resource management approach, 19–21
Holmgren, David, 45
Honey (family cow), 26–27, 158
honey bees
    beekeeping practices, 206–9
    ccolitcracy about, 232–34
    stresses on, 194–95
honeyberries
    growing information, 142–44, *143*
    pollen and nectar from, 203
    propagation of, 149
Honeycrisp apple, *93*
honey locust, 63, 84, 95
hoophouses, *13*
    addition to farm, 48–49
    crops grown in, 14
    frost damage prevention with, 114
    maintenance of, 92
    plum production, 14, 92, *113*
    pollination via bumble bees, 209–10
    raspberry production, 11, 14, 49
    repurposing, 88–94, *89*, *90*, *93*
    turkey production, *24*
hoophouse-style chicken tractors, 22, *23*
Hope (calf), 26–27
horses
    black locust toxicity, 163
    draft horses, 30–32, *32*, 64
    grazing areas, 170–71
    Knoll Orchard silvopasture, 99–100, 102
hügelkultur (mound culture)

bumblekulturs, 9, 86, 199, *199*
    climate change adaptations, 85–86, *85*
    cover crops for, 71
    perennial plantings in, 9
humans
    as ecosystems, 234
    as invasive species, 182–83
Hurricane Irene (2011), 7, *72*, 75

I

imported currant worm. *See* gooseberry sawfly
India, farmer-to-farmer exchanges, 230
industrial beekeeping, 207–8
industrial organic farming, 42
insects
    beneficial, 187, 188
    food web role, 152
    learning to love, 236–39, *236*, *239*
    in the pepinyè garden, 155, 156
    return of, through rewilding endeavors, 49–50
    See also specific types
integrated pest management (IPM), 117
invasive species. *See* pests and invasive species
ironweed, 165, *239*
irrigation
    dry-weather watering, 86–87, *87*
    minimal hoophouse needs, 92, 94
    for new perennial plantings, 106
Izzo, Vic, 56

J

jams, 137, 143
Jane (farmer), 33
Japanese-American hybrid plums, 124
Japanese beetles, 179–181, *180*

Index

Jerusalem artichokes.
    See sunchokes
Joan J raspberry, 88
Johns elderberries, 140, 142
Juneberry, 147*t*

K
kestrels, 174
Knoll Orchard
    alley cropping, *104*
    Cornell soil health
        assessment, 59, 59*t*
    gradual approach to
        planting, 109
    heritage apples in, 15
    milkweed stands, 167
    no-till production in
        alleys, 80
    polyculture fruit trees and
        guild plants, 96–97, *98*
    silvopasture and alley
        cropping, 99–103, *100*,
        101, *102*
    soil of, 113–14
Kosher King cockerels, 23, *23*

L
labeling
    organic certification,
        42–43
    Regenerative Organic
        Certification add-on
        label, 43
labyrinth, willow, 166–67
lambs
    bottle-feeding, *18*
    on-farm slaughtering of, 29
Lamoille River floodplain, 74
Lamoille River valley, 2, *3*, 76
landscape fabric mulch
    alley cropping, 80
    elderberry plantings, *80*,
        82–83
    importance in no-till
        systems, 79, 84–85
    in the pollinator sanctuary,
        164, 172
    reusable nature of, 177
landscape memory, 2–4

landscaping plants
    clove currants, 133, *133*
    elderberry and aronia,
        139, *155*
La Vía Campesina, 45
lawn orchard. *See* Front Lawn
    Orchard
lawns, reinvention into
    gardens and food forests, 54
laying hens, 24, *25*, 68–69
leaf-cutting bees, 154, 200,
    201, *201*
Liberty apples, *112*, *118*
limits to growth, 234–36
*Limits to Growth: The 30-Year
    Update* (Meadows, Randers,
    and Meadows), 235
lingonberries, 6, 51, 60, 147*t*
littleleaf linden trees, 151, 165
locust trees. *See* black locust
Louke, Joanna, 237

M
Magdoff, Fred, 60
Malling 7 rootstock, 115
management intensive grazing
    benefits for pasture,
        32–33
    early years, 21–27, *23*, *24*,
        25, *27*
    horses, 99–100, *102*
    in the pollinator sanctuary,
        170–71
    prior to polyculture
        plantings, 105
    sheep, *16*
manure
    chicken, 22, 33, 68
    cow, 24
    horse, 9, 24, 31, 64, 100
    rabbit, 33
Marge elderberries, 139–140,
    142, 186
marketing
    advertising, 227
    early years, 33
    names for the farm and
        related enterprises, 218,
        220–21, 222

organic certification
    and, 41–43
mason bees, *201*
mason wasps, 201–2, *201*
McKenzie aronia, 139
Meadows, Dennis, 235
Meadows, Donella, 235
meadows, in the pollinator
    sanctuary, 170–71, 204–5
meat CSAs, 19, 26, 33
meat eating, 29
    *See also specific types
        of livestock*
microorganisms
    contributions to soil
        biodiversity, 62–63
    importance to healthy
        soil, 57
    role in composting, 64,
        66–68
milking routine, family cows,
    26–27
milkweed
    alley cropping, 100, *102*
    growing information, 165,
        167–170, *168*, *170*
    in wildflower meadows, 15
milkweed floss, 169–170
milkweed seeds, 170, *170*
Minaj Smyriou black
    currant, 131
minor elements, for soil
    health, 60
mites, *Varroa*, 207, 208
monarch butterflies, 167–170,
    168, *239*
mound culture. *See* hügelkul-
    tur (mound culture)
Mount Mansfield, 2, *173*
mowing
    Front Lawn Orchard,
        119–120
    monarch butterfly
        concerns, 169
    orchard alleyways, 77, 84
    prior to polyculture
        plantings, 105
    reduced, as part of rewilding
        endeavors, 49, *52–53*, 54

mulberry trees, 125
mulching practices, for no-till production, 79, 80–85, *80, 81, 82–83*
   See also landscape fabric mulch

N

names for the farm and related enterprises, 218, 220–21, 222
Nanking cherry, 147t
nannyberry shrubs, 163–64
National Organic Program (NOP), 41–42, 43
native bees, 193–94, 195, 210
   See also specific types
native plantings
   creating niches for, 183
   in the pollinator sanctuary, 152, 171–72
   for protection of the seasonal stream, *8*
   rewilding endeavors, 15
   See also specific types
Nellie (horse), 31, *32*, 212
neonicotinoids, 195, 197
Nero aronia, 139
netting
   gooseberry sawfly larvae concerns, 189
   for protection against birds, 134, 143, 178
New York ironweed, 165
ninebark shrubs, 164
nitrifiers, 62–63
nitrogen-fixing bacteria, 62, 63, 70
nitrogen-fixing trees, 84, 98–99
NOP (USDA National Organic Program), 41–42, 43
Nora (horse), 31, *32*, 163
Northeastern Pollinator Plants, 172
Northeast Organic Farming Association (NOFA), 41, 212, 227
Northeast Sustainable Agriculture Research and Education (SARE), 48, 103, 131, 213
no-till production
   carbon sequestration benefits, 63
   farm's focus on, 7
   mulching practices, 79, 80–85, *80, 81, 82–83*
   pumpkins, 49
   regenerative no-till soil practices, 78–80, *79*
Nova elderberries, 140, 142
nursery area
   map of, *10, 11*
   retail nursery, 7, 9
   See also retail nursery
nursery plants
   fruits, 224–25, *225*
   propagation of, 146–49
   watering of, 86, 87
nut trees, 125–27, *126*
nylon footies for berry protection, *185*

O

oats, for cover cropping, 70–71, *71*
observation, importance of, 213
occultation, 84
off-farm income, 227–28
off-season resting time, 176, *176*
Old Home × Farmingdale, 97
   or 333 rootstock, 97
on-farm slaughtering, 28–29
open spaces, in the pollinator sanctuary, 170–71
organic certification, 41, 42–43
organic farming
   apples, 51, 116–17, 223
   fruit production, 51
   importance of soil health, 55
   industrial organic farming, 42
   overview, 39–43
   systems approach to, 19
organic matter
   complexity of, 61
   from cover crops, 69–70
   from decomposition of plant residues, 62
   importance in soil structure, 58, 61
   oxidation of, 63
   soil health assessment, 60–61
Orion, Tao, 183
ostrich ferns, 174
overshoot of populations, 234–35
owlet moths, 156

P

panicles, 139
parasitic wasps and flies, 188
pasture
   early years, 22
   management intensive grazing benefits, 32–33
   rewilding endeavors, 54
   silvopasture, 99–103, *100, 101, 102*
pasture-raised meat, benefits of, 29, 30
peaches
   growing information, 124
   in hoophouses, 14, 121
   self-fruitfulness of, 122
Pear Corner Orchard, 15, 96, 97, 121, 158–59
pears
   growing information, 120–21
   raised bed plantings of rootstock, 78
   rootstock for, 97
peas, for cover cropping, 70, 71, *71*
pepinyè garden, 6, *11*, 152–56, *154, 155*
perennial conservation mix plots, 205–6
perennial fruit production
   beginnings of, 48–49
   biodiversity benefits, *8*
   decision to pursue, 38
   farm's focus on, 7
   fields for, 9, *10*

Index

no-till practices, 78–80
polyculture fruit trees and guild plants, 96–99, *98*
See also specific crops
perennial shrubs, flood-tolerance of, 75
perennial vegetables, 103–5, *104*
permaculture overview, 45–46
Permaculture: Principles and Pathways beyond Sustainability (Holmgren), 45
perry (pear cider), 120
pesticide-free approach, support for bees, 195–97, *196*
See also organic farming
pesticides
    imbalance in predatory-prey relationships from, 190
    impact on bees, 195, 207
pest outbreaks, 178, 179
pests and invasive species, 177–192
    apple production, 51, 117–18, 186–87, *186*
    biodiversity benefits, 177–79, 180–81, 187
    bishop's-weed, 191–92
    elderberry production, 141
    fences and screens for, *104*
    gooseberry sawfly, 188–191, *189*
    hoophouse benefits, 14
    humans, 182–83
    industrial organic farming practices, 42
    integrated pest management, 117
    Japanese beetles, 179–181, *180*
    pests vs. pest outbreaks, 178, 179
    spotted wing drosophila, 91, 141, 183–86, *185*
    weed management, 177–78
    wild parsnip, 181–82
Peter (employee), *32*
petting pens, 214–15
Petunia (potbellied pig), 213–14, 215

phenology, 1
phoresy, 67
phosphorus, for soil health, 60
photosynthesis, 62
physical indicators of soil health, 59–60, *59t*
phytophotodermatitis, 182
piano, as wildlife habitat, *156*, 157–58
pigs
    early years, 26
    hogging down of cover crops, *27*
    slaughtering of, 28–29
plasterer bees, 202
plastic wrap guards, 107, 122
Plum Alley, 96
plum curculio, 187
plums
    American plums, 115, 124, *124*, 153, 156
    beach plums, *12–13*, 144–46, *145*
    first plantings of, 121, 122
    in hoophouses, 14, 92, *113*
    rootstock for, 78, 115
Polana raspberry, 88
pollination
    floral resources for, 203–6, *204t*, *206*
    importance of native bees, 193–94
    raspberries, 91
    See also specific types of pollinators
pollinator conservation projects, 230
pollinator pathway, 151–52, 171, 175, 176
pollinator sanctuary, 151–176
    black walnut plantings, 126–27
    deer protection in, 106–7
    development of, 14–15
    floral resources for bees in, 204–5
    hedgerow in, 95–96
    honey bee visiting willow flower, *150*

milkweed stands, 165, 167–170, *168*, *170*
native grass plantings, 171–72
off-season resting time, 176, *176*
open spaces and meadows, 170–71
overview, 151–52
pepinyè garden, 6, *11*, 151–52, *154*, *155*
reclaimed grass areas, 163–65
riparian zones and wildlife corridors, 156–58, *156*
stone fences, 174–75
tree plantings, 158–163, *160–61*, 172–74, *173*
willow labyrinth, 166–67
"Pollinators in Peril" (fiber arts series), *233*
polyculture systems, 105–8, *108*
population overshoot, 234–35
potassium, for soil health, 60
The Power of Restraint (Rabhi), 235
precautionary principle, 197
propagation
    grafting, 114–16, *116*
    uncommon fruits, 146–49, *146*
propagation beds, *146*, 148
pruning
    bumblekulturs from prunings, 9, 86, 199, *199*
    fruit and nut trees, 113
Prunus mahaleb rootstock, 115
pseudoscorpions, 67–68
pumpkins
    alley cropping, *101*, 102
    as living mulch, 84
    no-till production, 49, 80
pussy willow, 78, 163–64, 203

R

Raap, Will, 230
rabbits, 70, *71*
Rabhi, Pierre, 235

rainwater
    collection and storage of, 86–87, *87*
    soil infiltration, 2
raised beds, 78
    See also hügelkultur (mound culture)
Ranch elderberries, 140
Randers, Jorgen, 235
raspberries
    in hoophouses, 11, 14, 49, 88–89, *89*, 91, 92, 94
    nesting bees in old canes, 202
    picking hygiene for managing spotted wing drosophila infestations, 184–85
rats, in the barn, 69
Real Organic add-on label, 43
red currants, 134–35, *134*, 188–191, *189*
red-humped caterpillars, *186*
red osier dogwood, 154, 156
reed canary grass, 22, 163–65, 172
reforestation of Vermont, 3–4
regenerative agriculture
    farm's focus on, 7
    overview, 43–44
    See also specific practices
regenerative no-till soil practices, 78–80, *79*
"Regenerative Organic Agriculture and Climate Change" (Rodale Institute), 44
Regenerative Organic Certification add-on label, 43
resilience
    need for changes after the early years, 14, 32–35, 37–38
    overview, 38–39
    stressors and corresponding resilience strategies, 40–41*t*
    See also climate change adaptations

retail nursery, 7, 9
    See also nursery area
rewilding endeavors, 49–54, *50*, *52–53*, *54*
    See also pollinator sanctuary
Rhode Island Red laying hens, 24
rhubarb, *12–13*
    in field 2, 9, 11
    growing with asparagus, 103, *104*
    mulch from leaves, 84
    wholesale markets, 226
riparian zones
    bird and pollinator habitat in, 95, *96*
    climate change adaptations, 76–77, *82–83*
    elderberry plantings, 142
    pollinator sanctuary, 156–58, *156*
River Berry Farm, 33
Rodale Institute, 43, 44
rolling and crimping, 78
root rot (*Armillaria mellea*), 125
rootstocks
    apple trees, 97, 115
    grafting and, 114–16, *116*
    pear trees, 97
    plum trees, 78, 115
Roundup (glyphosate), 63
Rovada red currants, *134*
row plantings, 105, *106*
Rural Vermont, 212, 227

S

safety concerns
    black locust trees, 163
    elderberries, 142
    wild parsnip juice, 182
Salatin, Joel, 19, 22
Sarah (farm worker), *80*, 132
SARE (Northeast Sustainable Agriculture Research and Education), 48, 103, 131, 213
Saskatoon berry, 147*t*
sawfly, gooseberry, 188–191, *189*

scab, apple, 118
schist bedrock, 2
scion wood, 114–16, *116*, 146
scything, 107–8, *108*, 120
seaberries, 147*t*
sea buckthorn, 147*t*
sea kale, 102, 103, *104*
seasonal waterways, 2–3
Seeds of Self-Reliance, 230
seeps, 2–3
self-fruitful stone fruits, 122
semidwarf rootstocks, 115
sense of place, 1
serviceberries, 203
sharing the farm, 211–230
    agritourism programs, 213–15, *214*, *216–17*
    decisions on product expansion, 221
    farmer-to-farmer exchanges, 87, 228–230, *229*
    fruit ciders, 222–24, *223*
    fruit nursery, 224–25, *225*
    fruit syrups and snow cones, 218–221, *219*, *220*
    importance of flexibility, 226–27
    name choice for farm, 218, 220–21
    off-farm income, 227–28
    student programs, 211–13, *212*
    wholesale fruit sales, 226
sheep
    on-farm slaughtering of, 29
    lambs, *18*, 29
    management intensive grazing, *16*
sheet mulching, 79, 80–81, *106*, 177
Siberian pea shrub, 63, 84
Siberian quill, 203
silky dogwood, 154, 155, 156
silos, 7
silver maple
    in the pollinator sanctuary, 163–64
    woodchips from, 78, 81, 84

Index

silvopasture, 99–103, *100, 101, 102*
simplicity, 235–36
slaughtering animals, 28–30
small things, power of, 240
snow cones and fruit syrups, 218–221, *219, 220*
social resilience, 39
soil, 55–71
    biodiversity in, 62–63
    composting benefits, 9, 63–69, *65, 67*
    cover-cropping benefits, 69–71, *71*
    early years, 22
    factors in soil health, 58–62, 59*t*
    fruit and nut tree needs, 112, 113–14, 121
    living nature of, 56–58, *56*
    management intensive grazing benefits, 33
    types of, on the farm, 2
    wet and heavy, 77–78
soil pH, 60
soil respiration, 60, 62
soil structure
    damage to, from tillage, 7, 63, *79*
    organic matter for, 58, 61
    wear and tear of the early years, 34
soil testing
    comprehensive assessment test, 58–60, 59*t*
    early years, 22
solarization of damaged berries, 184–85
solitary nesting bees, 200–203, *201, 202*
southwest disease, 122
speckled alder, 84
spinosad (Entrust), 190
spot pickers, 132
spotted wing drosophila (SWD), 91, 141, 183–86, *185*
spring bulbs, pollen and nectar from, 203
squash bees, 102, 197, 200, 202

standard rootstocks, 115
steam juicing, 219
Steiner, Rudolf, 46
stereomicroscopy, 67
stocking density, 21
stone fences, 174–75
stone fruits
    favorite varieties, 123*t*
    growing information, 121–24
    in hoophouses, 92
    See also specific crops
stooling, 114–15
strawberries, spotted wing drosophila infestations, 184
strawberry-rhubarb syrup, 219, 220
Streamco willow, 153
strigs, 134, 135
student programs at the farm, 211–13, *212*
subsurface hardness, of soil, 59–60
suckering plants, 126
sugar maple trees, 159
sunchokes
    alley cropping, 102, 103, *104*, 105
    for mulch, 84
sunflower bees, 102
sunflowers
    alley cropping, *101*, 102
    no-till production, 80
surface hardness, of soil, 59–60
Sustainable Agriculture Research and Education (SARE), 48, 103, 131, 213
swamp milkweed, 165
swamp white oaks, 166
SWD. *See* spotted wing drosophila (SWD)
sweet cherries, 122

T

tachinid fly, 181
Tallamy, Doug, 152, 225
tart cherries
    first plantings of, 121, 122

    growing information, 122
    rootstock for, 115
thermophilic composting, 64–66, *65*
thinning, of apples, 117
tillage, negative effects of, 7, 63, *79*
    See also no-till production
tillering, avoiding, 78
Tillers International, 30
tipping and tailing, gooseberry tips, 137
Titania black currant, 131
tomatoes
    in hoophouses, 48, *49*, 88, 91
    pollen and nectar from, 203
topworking of trees, 115, *116*
tractor use, 64, 65, 66
tree plantings, in the pollinator sanctuary, 158–163, *160–61*, 172–74
    See also specific types of trees
tree tubes, 148
trunk wraps, 107, 122
Tundra honeyberries, 144
turkeys, *24*, 33
tussock moth caterpillars, 169

U

uncommon fruits, 129–149
    aronia, *54, 76*, 138–142, *138*, 148–49, 226
    beach plums, *12–13*, 144–46, *145*
    currants overview, 129–135, *133, 134*, 136*t*
    elderberries overview, 138–142, *140*, 141*t*
    gooseberries, 78, 136–38, 136*t*, 146, 148, *220*
    honeyberries, 142–44, *143*, 149, 203
    miscellaneous varieties grown on the farm, 147*t*
    propagation of, 146–49, *146*

257

US Department of Agriculture (USDA)
- National Organic Program, 41–42, 43
- relaxing of organic rules, 43
- Specialty Crop Block Grant program, 209–10

V

Valley Forge elm trees, 162
Varroa mites, 207, 208
vegetable production
- harm to soil from, 78, *79*
- perennial vegetables, 103–5, *104*
- stress of annual tasks, 73
- *See also specific crops*

vermicomposting, 66, *67*
Vermont
- climate change effects, 91–92, 159
- history of, 3–4

Vermont Farm to Plate, 226
Vermont Farm Viability Enhancement Program, 227
Vermont Organic Farmers (VOF), 41
Vermont Sustainable Jobs Fund, 226
vernal pools, 15, 175
La Vía Campesina, 45
viceroy butterflies, 169
Viking aronia, 139
Virginia rose, 154, 163
VOF (Vermont Organic Farmers), 41
The Voice of the Infinite in the Small (Louke), 237
voles
- landscape fabric considerations, 84, 177–78
- protection against, *104*, 107, 108, 122

Voluntary Simplicity (Elgin), 235

W

wabi sabi (Japanese concept), 47
Warner, Chauncey, 175
The War on Bugs (Allen), 237
wasps
- bee boxes for, 201–2, *201*
- ecoliteracy about, 232–34
- in hoophouses, 89, 91

water harvesting and storage, 86–87, *87*
waterways, seasonal, 2–3
Watts, Alan, 18
weed management, 177–78
Whatley, Booker T., 19
whip and tongue grafting, 115
white-pine blister rust, 130–31
wholesale fruit sales, 48, 226
wildlife corridors, 156–58, *156*
wildness
- benefits of wild places, 15
- farm's focus on, 7
- *See also* rewilding endeavors

wild parsnip, 181–82
Wild Plum Cider, 146
willow
- hardwood cuttings of, *146*
- honey bee visiting flower of, *150*
- in the pepinyè garden, 153
- pollen from, for bees, 203
- in the pollinator sanctuary, 163–64
- shrub plantings, 78
- woodchips from, 81, 84

willow labyrinth, 166–67
windbreaks, hedgerows as, 95–96, 126, 154, 205
wineries, wholesale markets for fruit, 14, 130, 226
winterberry holly, 166
wintergreen, 147*t*
winter rye cover crops, 70, 78
witch hazel, 155–56
wolf willow plantings, 99
woodchips
- mulch from, 79, 80, 81, *81*, 84
- plantings for, 78

wrap guards, 107, 122
Wyldewood elderberries, 140

Y

yellowjackets (wasps), 232
yellow-twig dogwood, 154
York elderberries, 140, 142

About the Authors

JESSICA SIPE

Nancy J. Hayden is a writer, farmer, artist, and former environmental engineering professor. She's earned degrees in biology and ecology, environmental engineering, English, studio art, and creative writing. She was awarded a Vermont Arts Council Creation Grant to work on this book and has published numerous articles about food and farming. A keen student of World War I history, she recently published *The Great Dark: Noir and Horror Stories of World War One*. Her writing website is www.nancyjhayden.com.

John P. Hayden has been working to design and implement agricultural systems with positive environmental and social outcomes for over thirty-five years as a researcher, extension agent, university educator, international consultant, and practicing regenerative organic farmer. His farming and business experience include organic livestock, vegetables, fruit and nursery production, and marketing. He has an MS in entomology with a focus on ecological pest management and has served on Vermont's Pollinator Protection Committee and several nonprofit boards.

Their farm website is www.thefarmbetween.com.

the politics and practice of sustainable living
CHELSEA GREEN PUBLISHING

Chelsea Green Publishing sees books as tools for effecting cultural change and seeks to empower citizens to participate in reclaiming our global commons and become its impassioned stewards. If you enjoyed *Farming on the Wild Side*, please consider these other great books related to biodiversity and regenerative agriculture.

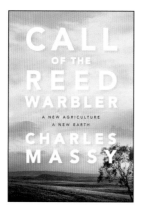

CALL OF THE REED WARBLER
A New Agriculture, A New Earth
CHARLES MASSY
9781603588133
Paperback • $24.95

DANCING WITH BEES
A Journey Back to Nature
BRIGIT STRAWBRIDGE HOWARD
9781603588485
Hardcover • $24.95

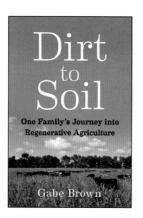

DIRT TO SOIL
One Family's Journey into Regenerative Agriculture
GABE BROWN
9781603587631
Paperback • $19.95

TOP-BAR BEEKEEPING
Organic Practices for Honeybee Health
LES CROWDER and HEATHER HARRELL
9781603584616
Paperback • $24.95

For more information or to request a catalog, visit **www.chelseagreen.com** or call toll-free **(800) 639-4099**.